The Hunting Farmers

Understanding ancient human subsistence in the central part of the Korean peninsula during the Late Holocene

Seungki Kwak

Access Archaeology

Archaeopress Publishing Ltd
Gordon House
276 Banbury Road
Oxford OX2 7ED

www.archaeopress.com

ISBN 978 1 78491 675 6
ISBN 978 1 78491 676 3 (e-Pdf)

© Archaeopress and S Kwak 2017

Printed and bound in Great Britain by
Marston Book Services Ltd, Oxfordshire

All rights reserved. No part of this book may be reproduced or transmitted,
in any form or by any means, electronic, mechanical, photocopying or otherwise,
without the prior written permission of the copyright owners.

Contents

Foreword xi

1. Subsistence change, Emergence of agriculture, and Rice 1

1.1. The role of the intensive rice agriculture in the central part of the Korean Peninsula 1

1.2. The transition from foraging to farming and the emergence of agriculture 2

1.3. What do we know so far about transitions to agriculture? 4

    1.3.1. Migrations of farmers 4

    1.3.2. Genetic studies 4

    1.3.3. First contact 5

1.4. Cases in East Asia 5

    1.4.1. China 5

    1.4.2. Japanese Archipelago 5

2. Background and Central Hypothesis 7

2.1. Archaeology in Korea - Its brief history and social context 7

2.2. Chulmun foragers and Mumun farmers - where everything started 8

2.3. Current views on the transition from foragers to farmers and development of rice agriculture in the Korean Peninsula 12

2.4. The central hypothesis of this study 14

2.5. The Chulmun and Mumun periods: Essentialism vs. Materialism 15

3. Methodological background, Research design and analytical procedure of the Luminescence dating 19

3.1. Luminescence dating in archaeology 19

3.2. Luminescence: The principals 20

    3.2.1. Thermoluminescence 21

    3.2.2. Optically stimulated luminescence 21

3.3. Limits of the luminescence dating 22

    3.3.1. Resetting of the signal 22

    3.3.2. Accuracy and precision 22

    3.3.3. Upper and Lower age limits 22

3.4. Luminescence dating and its application to the Korea archaeology 23

3.5. Analytical procedure 24

    3.5.1. Sample preparation - grain size 24

    3.5.2. Glassware and reagents 25

    3.5.3. Dose rate measurement 25

3.5.4. Equivalent dose measurements 25
3.5.5. Determining the age 26
4. Methods, Research design and analytical procedure of the organic geochemical analysis 27
4.1. Concept of biomolecular archaeology and organic geochemical analysis 27
4.2. Organic Residues within archaeological potteries 27
4.3. Identification of lipids 30
    4.3.1. GC-MS analysis 30
    4.3.2. Compound specific isotope analysis 31
        4.3.2.1. Modern reference animal fats and plant oils 34
        4.3.2.2. Interpretation of CSIA 35
        4.3.2.3. Possibilities of variation in $\delta^{13}C$ values of the fatty acids from the archaeological lipid 36
            4.3.2.3.1. Forest density and depletion of ^{13}C 36
            4.3.2.3.2. Variations in $\delta^{13}C$ values of CO_2 37
            4.3.2.3.3. Variation related to human activities 37
4.4. Analytical procedures 39
    4.4.1. Glassware, solvents and reagents 40
    4.4.2. Solvent extraction of lipids 40
        4.4.2.1. Preparation of TMS derivatives 41
        4.4.2.2. Preparation of FAMEs 41
    4.4.3. Methanolic acid extraction of lipids 42
    4.4.4. Analysis with GC-MS and GC-C-IRMS 42
        4.4.4.1. High Temperature GC-MS 42
        4.4.4.2. GC-C-IRMS 42
5. The Results 44
5.1. Kimpo-Yangchon 44
    5.1.1. Sampling 46
        5.1.1.1. Organic geochemical analysis 46
        5.1.1.2. Luminescence dating 51
    5.1.2. Organic geochemical results 51
    5.1.3. Luminescence dating results 56
5.2. Sosa-Dong 56
    5.2.1. Sampling 60
        5.2.1.1. Organic geochemical analysis 60

    5.2.1.2. Luminescence dating...61

    5.2.2. Organic geochemical results..61

    5.2.3. Luminescence dating results...64

5.3. Songguk-Ri...68

    5.3.1. Sampling...70

        5.3.1.1. Organic geochemical analysis...70

        5.3.1.2. Luminescence dating...71

    5.3.2. Organic geochemical results..71

5.4. Eupha-Ri..76

    5.4.1. Sampling...79

        5.4.1.1. Organic geochemical analysis...79

        5.4.1.2. Luminescence dating...79

    5.4.2. Organic geochemical results..80

    5.4.3. Luminescence dating results...85

5.5. Isotope implication: Alternative approach 85

6. Discussion...90

6.1. The subsistence of the Chulmun and Mumun periods...90

    6.1.1. The Chulmun subsistence..90

    6.1.2. The Mumun subsistence 93

6.2. The subsistence of Iron Age..94

6.3. Luminescence dating..95

6.4. Implications and future directions 96

7. Conclusion..99

7.1. Reprising the work so far..99

7.2. Transition from foraging to farming: Theoretical model vs. Empirical world 100

7.3. Concluding remark: The role of intensive agriculture as a subsistence strategy in the prehistoric Korean peninsula..101

8. Bibliography...103

List of Figures

Figure 1.1 The map of the Korean Peninsula and its vicinity 6

Figure 2.1 (a) The indication of the central part of the Korean Peninsula (b) The location of the Konam-Ri shell midden 9

Figure 2.2 The Chulmun and the Mumun period potteries (a): a Chulmun pot with the combshape pattern and pointed bottom (b): two pieces of the Mumun pottery with patterns mostly on the rim (upper-right; Heunam-ri-style) and with no pattern (downright; Songguk-Ri-style) (modified from Yoon & Bae, 2010) 10

Figure 2.3 The Chulmun and Mumun period tools (modified from Yoon & Bae, 2010). 11

Figure 2.4 The patterns on the Mumun pottery (a): pinched strip (Cheon, 2005) (b): Garak-Dong style (B. G. Lee, 1974) (c): Yeoksam-Dong style (B. G. Lee, 1974) (d): Heunam-Ri style (J. H. Ahn, 2000) (e): Songguk-Ri style (Norton, 2007) 11

Figure 2.5 The Chulmun and Mumun period habitations (modified from Yoon & Bae, 2010) 12

Figure 2.6 The evidence of the full-dress farming in the central part of the Korean Peninsula: (a) stone sickles, (b) semi-lunar shaped knives, (c) excavated dry field, and (d) irrigated rice paddy (all modified from Yoon & Bae, 2010) 14

Figure 2.7 Variation in pattern on the Chulmun potteries (a): Amsa-Dong (b): Amsa-Dong (c): Sammok island (d): Amsa-Dong (e): Yongyou island 16

Figure 2.8 Schematic representation of the essentialist ('typological' thinking) and materialist ('population' thinking) approaches (modified from Marwick, 2008: p. 108) 17

Figure 2.9 Location of the sites mentioned in the text 18

Figure 3.1 A typical thermoluminescence signal (commonly referred to as "glow curve") that shows multiple traps (Duller, 2008; cf. Feathers, 2003: 1495) 21

Figure 3.2 A typical optically stimulated luminescence signal from quartz grains (Duller, 2008) 22

Figure 4.1 partial HTGC profile of the lipid extract from a Romano-British sherd from Stanwick, Northamptonshire (Evershed et al., 2002: 661). A low abundance of intact TAGs are observed at retention times above 30 min. The majority of them was hydrolyzed during vessel use or burial, resulting in the formation of DAGs, MAGs, and free fatty acids. Key: IS = internal standard (n-tetratriacontane). IS was added to the sample at the extraction stage for quantification of lipid. The extracts are trimethylsilylated 30

Figure 4.2 The distributions of TAGs in different kinds of animal fats (modified from Mukherjee, 2004: p. 20). (a): cow adipose fat (b): sheep adipose fat (c): pig adipose fat (d): fresh milk (e): milk degraded for 90 days 32

Figure 4.3 Undiagnostic C16:0 and C18:0 fatty acids generated through the hydrolysis of triacylglycerols due to the degradation of fat/oil during burial process. As biomarkers, C16:0 and C18:0 fatty acids have a severely limited diagnostic value (Evershed, 2008b: 900) 33

Figure 4.4 Reference database created based on modern fats for CSIA. (a): Only the animals having been reared on known diets were sampled (e.g.C3 plant diet in order to mimic the prehistoric condition, absence of C4 plants) (Copley et al., 2003; Dudd et al., 1998). (b): The modern reference samples were collected from authentic wild animals to avoid the effects of commercial farming and selective breeding (Craig et al., 2013). The $\delta^{13}C$ values obtained from all modern reference animals were adjusted by the addition of 1.2 permil, considering post-Industrial Revolution effects of fossil fuel burning (Friedli et al., 1986) 35

Figure 4.5 Interpretation of the results of CSIA (a): $\delta^{13}C$ values acquired from the C16:0 and C18:0 fatty acids in archaeological potsherds are plotted along with the reference animal fat ellipses (Evershed, 2007) (b): The theoretical mixing curves between the porcine adipose fat, ruminant adipose fat and ruminant dairy fat are shown (Evershed, 2008b: 900) 36

Figure 4.6 Plots showing the difference in δ13C values of the C18:0 and C16:0 fatty acids ($\Delta^{13}C$ = $\delta^{13}C18:0$ - $\delta^{13}C16:0$) obtained from the modern reference fats (Copley et al., 2003: 1526) 38

Figure 4.7 $\delta^{13}C$ values of the cellulose from the oak tree-ring sequence (a): 11-year running mean from ancient Irish oaks (data obtained from McCormac et al., 1994; Mukherjee, 2004: 28) (b): Yearly measurement from 1970 to 1995 of modern oaks in east England (Robertson et al., 1997) 39

Figure 4.8 The comparison between (a) the solvent extraction protocol and (b) the acid extraction protocol (Correa-Ascencio & Evershed, 2014: 1331) 41

Figure 4.9 The GC chromatograms of the same archaeological sherd sample (KIM014) showing different recovery rates. (a) chloroform : methanol solvent extraction (b): acidified methanol extraction (IS = Internal Standard). In both extractions, the same amount of internal standard was injected. The acidified extraction method showed a much higher recovery rate (more than 20 times) compared with the prevailing chloroform /methanol solvent extraction protocol... 43

Figure 5.1 The density distribution of radiocarbon the dates from the Kimpo-Yangchon site, using the R package BChron (the dates were calibrated using the "intcal13" calibration curve, cf. Reimer et al., 2013) .. 46

Figure 5.2 The location of the four sites analyzed in this thesis 47

Figure 5.3 The site plan of the Kimpo-Yangchon site and the location/number of the samples taken for the radiocarbon dating (R), organic geochemical analysis (O), and luminescence dating (L) (B. M. Kim et al., 2013) 48

Figure 5.4 Some of the artifacts uncovered during the excavation of the Kimpo-Yangchon site: semi-lunar shaped knife (upper-left), pots (right; rim-punctuation: upper-right), and arrowheads (down-left) 49

Figure 5.5 The result chromatogram of the GC-MS analysis of one of the samples from the Kimpo-Yangchon site (KIM061), using R version 3.2.0. Due to degradation, we usually observe medium- and long-chain saturated fatty acids. 5-α Cholestane was added as an internal standard (IS = 132 ng / microliter... 52

Figure 5.6 The results of CSIA by GC-C-IRMS of the samples from the Kimpo-Yangchon site using the available references (cf. Dudd & Evershed, 1998; Dudd et al., 1999; Steele et al., 2010) 53

Figure 5.7 The results of CSIA by GC-C-IRMS of the samples from the Kimpo-Yangchon site using the reference from Craig et al. (2011) 54

Figure 5.8 The results of CSIA by GC-C-IRMS of the samples from the Kimpo-Yangchon site using the reference from Craig et al. (2013) 55

Figure 5.9 Some of the artifacts uncovered during the excavation of the Sosa-Dong site including potsherd, arrowheads and stone chisel. The potsherd in the picture has the rimpunctuation/short slanted incision... 58

Figure 5.10 (a): The carbonized rice grains (*Oryza sativa*) and (b): possible barley (*Hodeum vulgare* L.) grains excavated in the Area "Ga" house pit No. 10 (B. M. Kim et al., 2008) 59

Figure 5.11 The density distribution of the radiocarbon dates from the Sosa-Dong site, using the R package BChron (the dates were calibrated using the "intcal13" calibration curve, cf. Reimer et al., 2013) 59

Figure 5.12 Diagram showing the lipid concentration of each body part from the both experimental and archaeological sherd samples (adapted from Evershed, 2008a: 32) 60

Figure 5.13 The site plan of the Sosa-Dong site and the location of the samples taken for the radiocarbon dating (R), organic geochemical analysis (O), and luminescence dating (L) (B. M. Kim et al., 2008) .. 62

Figure 5.14 The result chromatogram of the GC-MS analysis of one of the samples from the Sosa-Dong site (SOS049), using R version 3.2.0. Due to degradation, we usually observe medium- and long-chain saturated fatty acids. 5-α Cholestane was added as an internal standard (IS = 132 ng / microliter) 63

Figure 5.15 The results of CSIA by GC-C-IRMS of the samples from the Sosa-Dong site using the available references (cf. Dudd & Evershed, 1998; Dudd et al., 1999; Steele et al., 2010) 65

Figure 5.16 The results of CSIA by GC-C-IRMS of the samples from the Sosa-Dong site using the reference from Craig et al. (2011) 66

Figure 5.17 The results of CSIA by GC-C-IRMS of the samples from the Sosa-Dong site using the reference from Craig et al. (2013) 67

Figure 5.18 (a): some of the artifacts uncovered during the excavation of the Songguk-Ri site: pot, large tubular-shaped greenstone ornaments, semi-lunar shaped stone knife, arrowheads, ground stone dagger, and Liaoning-style bronze dagger (Yoon & Bae, 2010) (b): the "Songguk-Ri style" rounded pit-house with two post holes (Yoon & Bae, 2010) 68

Figure 5.19 The density distribution of radiocarbon dates from the Songguk-Ri site, using the R package BChron (the dates were calibrated using the "intcal13" calibration curve, cf. Reimer et al., 2013).... 69

Figure 5.20 The site plan of the Songguk-Ri site and the location of the samples taken for the radiocarbon dating (R), organic geochemical analysis (O), and luminescence dating (L) (G. T. Kim et al., 2013) 72

Figure 5.21 The result chromatogram of the GC-MS analysis of one of the samples from the Songguk-Ri site (SON024), using R version 3.2.0. Due to degradation, we usually observe medium- and long-chain saturated fatty acids. 5-α Cholestane was added as an internal standard (IS = 123.2 ng / microliter) (Kwak et al., 2017: 8) 73

Figure 5.22 The results of CSIA by GC-C-IRMS of the samples from the Songguk-Ri site using the available references (cf. Dudd & Evershed, 1998; Dudd et al., 1999; Steele et al., 2010) 74

Figure 5.23 The results of CSIA by GC-C-IRMS of the samples from the Songguk-Ri site using the reference from Craig et al. (2011) 75

Figure 5.24 The results of CSIA by GC-C-IRMS of the samples from the Songguk-Ri site using the reference from Craig et al. (2013) 76

Figure 5.25 Lu (呂) shape and (b): Tu (凸) shape house pits excavated from the Eupha-Ri site (Wang et al., 2013) 77

Figure 5.26 The density distribution of radiocarbon dates from the Eupha-Ri site, using the R package BChron (the dates were calibrated using the "intcal13" calibration curve, cf. Reimer et al., 2013).... 78

Figure 5.27 Some of the artifacts uncovered during the excavation of the Eupha-Ri site (a): the Iron Age style hardened un-patterned pottery, a pot made by the beating method (center, second row) (b): mold for iron casting, net sinker, spindle whorls, iron axes and arrowheads 78

Figure 5.28 The site plan of the Eupha-Ri site and the location of the samples taken for the radiocarbon dating (R), organic geochemical analysis (O), and luminescence dating (L) (H. J. Wang et al., 2013) 80

Figure 5.29: The result chromatogram of the GC-MS analysis of one of the samples from the Eupha-Ri site (EUP005), using R version 3.2.0. Due to degradation, we usually observe medium- and long-chain saturated fatty acids. 5-α Cholestane was added as an internal standard (IS = 44 ng / microliter) 81

Figure 5.30 The results of CSIA by GC-C-IRMS of the samples from the Eupha-Ri site using the available references (cf. Dudd & Evershed, 1998; Dudd et al., 1999; Steele et al., 2010) 82

Figure 5.31 The results of CSIA by GC-C-IRMS of the samples from the Eupha-Ri site using the reference from Craig et al. (2011) 83

Figure 5.32 The results of CSIA by GC-C-IRMS of the samples from the Eupha-Ri site using the reference from Craig et al. (2013) 84

Figure 5.33: (a, b) The results CSIA of the samples from the Kimpo-Yangchon and Sosa-Dong site using the approach of Salque et al. (2013). (a) The $\delta^{13}C$ values suggest a C_3 biased across the samples, indicating terrestrial herbivores mainly consumed indigenous wild C_3 plants (cf. Ahn, 2006; Choy and Richards, 2010; J. J. Lee, 2011b). $\Delta^{13}C$ 18:0 = $\delta^{13}C$18:0 - $\delta^{13}C$16:0, KI = Kimpo-Yangchon, SO = Sosa-Dong 87

Figure 5.34: (a, b) The results CSIA of the samples from the Songguk-Ri site using the approach of Salque et al. (2013) (Kwak et al., 2017: 9). (a) The $\delta^{13}C$ values suggest a C_3 biased across the samples, indicating terrestrial herbivores mainly consumed indigenous wild C_3 plants (cf. Ahn, 2006; Choy and Richards, 2010; J. J. Lee, 2011b). $\Delta^{13}C$ 18:0 = $\delta^{13}C$18:0 - $\delta^{13}C$16:0, SO = Songguk-Ri 88

Figure 5.35: (a, b) The results CSIA of the samples from the Eupha-Ri site using the approach of Salque et al. (2013). (a) The $\delta^{13}C$ values suggest a C_3 biased across the samples, indicating terrestrial herbivores mainly consumed indigenous wild C_3 plants (cf. Ahn, 2006; Choy and Richards, 2010; J. J. Lee, 2011a; 2011b). $\Delta^{13}C$ 18:0 = $\delta^{13}C_{18:0}$ - $\delta^{13}C_{16:0}$, EU = Eupha-Ri 89

Figure 6.1 The results of the bulk isotope analysis on human remains and animal bones excavated from the Daepo and Tongsam-Dong shell middens (modified from J.J. Lee 2011b: p. 41) 91

Figure 6.2 The results of the bulk isotope analysis on human remains and animal bones excavated from the Konam-Ri shell middens (modified from J. J. Lee, 2011b: p. 44) 92

Figure 6.3 Density distributions of all radiocarbon dates from each site studied in this thesis, using the R package BChron (Sosa-Dong: SS, Kimpo-Yangchon: KM, Songguk-Ri: SG, Eupha-Ri: EP) All dates were calibrated using 'intcal13' calibration curve 92

Figure 6.4 The comparison between the AMS radiocarbon dates and luminescence dates of the four sites 96

List of Tables

Table 3.1 The result of the luminescence dating on the proto-historic period potsherd (*The dose rates are rounded to two decimal places, but the calculation of the total dose rate was carried out prior to rounding) 23

Table 4.1 Criteria used to distinguish food types, based on fatty acid ratios (Eerkens 2005) 29

Table 4.2 Identification of fatty acids by using GC-MS (Stear 2008: 26) 29

Table 5.1 The results of the AMS radiocarbon dating of the Kimpo-Yangcho site 45

Table 5.2 The samples collected from the Kimpo-Yangchon site for the organic geochemical analysis 50

Table 5.3 The samples collected from the Kimpo-Yangchon site for the luminescence dating 51

Table 5.4 The results of the luminescence dating of the potsherd samples from the Kimpo-Yangchon site 56

Table 5.5 The results of AMS radiocarbon dating of the Sosa-Dong site 57

Table 5.6 The samples collected from the Sosa-Dong site for the organic geochemical analysis 60-61

Table 5.7 The samples collected from the Sosa-Dong site for the luminescence dating 61

Table 5.8 The results of the luminescence dating of the potsherd samples from the Sosa-Dong site 64

Table 5.9 The results of the AMS radiocarbon dating of the Songguk-Ri site 69-70

Table 5.10 The samples collected from the Songguk-Ri site for the organic geochemical analysis in this study 70-71

Table 5.11 The results of AMS radiocarbon dating of the Eupha-Ri site 77

Table 5.12 The samples collected from the Eupha-Ri site for the organic geochemical analysis in this study 79

Table 5.13 The samples collected from the Eupha-Ri site for the luminescence dating in this study 80

Table 5.14 The results of the luminescence dating of the potsherd samples from the Eupha-Ri site. The overall low water content of the samples shows the less porous nature of the Iron Age pottery 85

Table 6.1 The comparison between the luminescence dates and AMS radiocarbon dates of the four sites 95

Foreword

The transition from foragers to farmers and the role of intensive rice agriculture have been among the most controversial subjects in Korean archaeology. However, the relatively high acidity of sediment in the Korean peninsula has made it impossible to examine faunal/floral remains directly for tracing the subsistence change. For this reason, many of the studies on the transition heavily relied on the shell middens in the coastal areas, which reflect only a small portion of the overall subsistence in the Korean Peninsula. The subsistence behaviors recorded in numerous large-scale inland habitation sites have been obscured by the overall separation between hunter-gatherer and intensive rice farmer. My dissertation research investigates the role of intensive rice farming as a subsistence strategy in the central part of the prehistoric Korean peninsula using organic geochemical analysis and luminescence dating on potsherds. The central hypothesis of this research is that there was a wide range of resource utilization along with rice farming around 3,400-2,600 BP. This hypothesis contrasts with prevailing rice-based models, where climatically driven intensive rice agriculture from 3,400 BP is thought to be the dominant subsistence strategy that drove social complexity. This research focuses on four large-scale inland habitation sites that contain abundant pottery collections to evaluate the central hypothesis as well the prevailing rice-centered model. This research produced critical data for addressing prehistoric subsistence of Korean peninsula and established detailed chronology of subsistence during 3,400-1,800 BP.

This monograph was supported by the Laboratory Program for Korean Studies through the Korean Studies Promotion Service/Academy of Korean Studies, Ministry of Education of Korea (AKS-2015-Lab-2250001). During my postdoctoral employment at the University of Oregon with the Laboratory Program, I was able to further elaborate my doctoral dissertation, entitled the "Long-term chronology of subsistence and the role of intensive agriculture in the central part of the Korean peninsula during the Late Holocene" for this book. The dissertation was a product that benefited by considerable advice and support from so many people. My academic advisor, Dr. Ben Marwick, provided me one of the rarest chance to study archaeology as a science. His passion, encouragement, and guidance had a vital role in the development of the thesis. My other committee members, Dr. Peter Lape and Dr. James Feathers gave me invaluable comments that made the final product incomparable to the first draft. The Graduate School Representative (GSR), Dr. Rick Keil, was as helpful as other committee members with his profound knowledge in organic chemistry. Considering the international and interdisciplinary character of the project, many collaborators and institutions from United States, United Kingdom, and South Korea assisted in the completion of the thesis. Dr. Julian Sachs, Dr. Joshua Gregersen, Dr. Daniel Nelson taught me how to prepare and analyze lipid samples. I was also truly benefited by the sincere help from Dr. Richard Evershed, Dr. Julie Dunne, and Dr. Marisol Correa-Ascencio at the Organic Geochemistry Unit, University of Bristol. I would not have gotten this far without their advice. Dr. Byeong-mo Kim (Korea Institute of Cultural Heritage), Dr. Gyeong-taek Kim (Korea National University of Cultural Heritage), Dr. Tae-seop Choi (Yonsei University), and Dr. Soojin Kong (Sejong University) kindly provided their Korean potsherd samples for this research. I would like to say special thanks to Ah-guan Kim (Korea Institute of Cultural Heritage) and Taehong Kang (Korea Institute of Cultural Heritage) for helping me collecting suitable samples. Also, thanks to Dr. Gyoung-Ah Lee (University of Oregon) and Dr. Chuntaek Seong (Kyung Hee University) for their sincere advice. In addition, the National Science Foundation DDIG grant provided funding during the doctoral research period.

During my six years of graduate student life, I was fortunate to be surrounded by fantastic colleagues at the University of Washington who helped so much in getting through the tough times. Anna Cohen, the best cohort colleague I have ever had, shared her broad knowledge in archaeology all the time. Erik Gjesfjeld, Jake Deppen, Natalia Slobodina, David Carlson, Rodrigo Solinis-Casparius, Joss Whittaker, Ian Kretzler, Jiun-Yu Liu, Lauryl Zenobi, Li-Ying Wang, and Gayoung Park were the best friends and colleagues at the same time. In addition to these amazing cohort, Catherine Ziegler and John Cady were so helpful for me to get through all the administrative complexities of the University.

Finally, none of these accomplishments would be possibly without sincere support from my friends and family. Dad, mom, and my sister, Joonku always encouraged me in my decision making, such as attending grad school. I would like to say special thanks to my father-in-law, mother-in-law, Kyeongin, Haerin, Peter, and Ray for their sincere support. To my best friends Heedong, Jaein, Kanghee, Soochul, Kihwan, Sangmin, Dongho, Jaesuk, Seowoo, Taeyoung, Kyujin, Youngjae, Seungoh, Changsoon, Moonsik, Jonghwan, and Keun-taek, you have been with me through and I can't wait to celebrate this accomplishment with you. Most importantly, I want to thank my wife for her years of sacrifice in allowing me to achieve this dream. Her devoted sacrifice is something that I may never be able to recompense, but I will try for the rest of my life.

1. Subsistence change, Emergence of agriculture, and Rice

1.1. The role of the intensive rice agriculture in the central part of the Korean Peninsula

According to the recent report from the Food and Agriculture Organization of the United Nations (FAO), the average annual rice consumption per person in Brunei and Vietnam is 245 Kg and 166 Kg (Faostat 2011). These two countries mark the 1st and 2nd in rice consumption in the world. The average annual rice consumption per person in South Korea in 2011 was 88 kg (the Korea National Statistical Office). However, according to historical records, the annual South Korean rice consumption per person around the 18th century was about 173 kg. Though the westernized life style of South Korea reduced its annual rice consumption rate, rice is still the mainstay of its modern diet, and has been so for at least 2,000 years. The Korean people's attachment to rice is remarkable. The word for 'meal' in Korean is 'bab', which also and originally means 'steamed rice'. Regardless of their economic status, way of life, or ideological inclination, steamed rice was and is the essential dish throughout the nation. For the Koreans, 'A bowl of rice is equivalent to love and affection' (Woo, 2012). In this regard, one of the main topics of Korean archaeology over the last 50 years has been investigating the process of the subsistence change from hunter-gatherers to intensive rice farmers. However, despite continuous attempts to reveal the overall pattern of the change and accumulations of data, we still lack information on some of the most basic parameters involved in the role of the intensive rice agriculture in the prehistoric Korean Peninsula.

The central part of the Korean Peninsula (Figure 1.1) contains a vast amount of archaeological data related to the subsistence change in the deeper past. This region has provided rich archaeological records documenting its general culture history. Its earliest known occupants were Paleolithic foragers dated to about 200,000 years ago (J. C. Kim et al., 2010). Clear evidence show that full-dress farming was practiced in this region around 3,400 BP (G. A. Lee, 2003, 2011). Solid evidence of dry fields, irrigated rice paddies and harvesting tools have been found (Yoon & Bae, 2010). However, due to the lack of paleobotanical evidence from this period, detailed information about when rice became the mainstay of the Korean diet is not yet known. Therefore, the study of the transition from hunter-gatherers to farmers, and the role of the intensive rice agriculture in this transition, is integral to anthropological debates.

The transition from foragers to farmers in the Korean Peninsula has been described as the subsistence change from hunter gathering to intensive rice farming around 3,400 BP (J. H. Ahn, 2000; B. C. Kim, 2006a, 2006b; J. S. Kim, 2003; Norton, 2000, 2007). B. Kim (2006a) argued that an agricultural economy based heavily on rice spread suddenly and swiftly into the foraging context with few evidences of a transitional period (cf. G. A. Lee, 2011). However, recent paleobotanical data on the southern part of the Korean Peninsula have revealed that people in this period were more dynamic and varied than is posited by the models focused on the intensive rice farming (Crawford and Lee 2003; G. Lee 2003, 2011). For example, along with rice, they utilized other crops such as millet, soybean, and azuki for their subsistence. These new data require an alternative model which could explain the role of the intensive rice agriculture in this period.

This monograph investigates the role of intensive rice agriculture as a subsistence strategy in the central part of the Korean Peninsula, contributing new data that helps to establish the chronology of subsistence over the last 3,400 years. This research will provide an insight into when rice became the mainstay of the Korean diet. Low hills with gentle slopes embracing meandering rivers in this region were continuously occupied for as much as 4,000 years, and large inland habitation sites developed in this condition provide the multiple lines of subsistence data that are required for this study. The central hypothesis in this research is that a wide range of resources were utilized along with rice between 3,400 and 2,000 BP. This hypothesis contrasts with the prevailing rice-centered models, which assume rice to be the most dominant subsistence resource since 3,400 BP.

The primary goal of this research is re-evaluating the conventional rice-centered models to better understand the overall pattern of subsistence strategy and assess the weight of rice in it. To achieve this goal the study (1) tests the hypothesis that a wide range of resources were utilized along with rice between 3,400 and 2,000 BP., and (2) establishes a general chronology of subsistence during this period, incorporating in that work the organic geochemical analysis and luminescence dating of the pottery excavated from four large inland habitation sites in the central part of the Korean Peninsula.

In Korean archaeology, pottery is one of the primary analytical resources, being abundant in almost every archaeological assemblage in the Korean Peninsula since 6,000 BP. However, despite intensive relative chronology-building, almost no attention has been given to analyzing the fabric of the pottery itself. Studies have showed that high-temperature boiling using pottery is particularly effective in the preparation of various resources (Stahl 1989; Wandsnider 1997). This represents a serious gap in our understanding of the prehistoric subsistence in Korea during the critical time of the transition from foragers to farmers. The methods proposed here allow me to test the prevailing rice-centered models, first by identifying what was stored and cooked in the pots, and second by dating the pots directly and absolutely. By doing so, the study establishes a general and robust chronology of subsistence between during 3,400 and 2,000 BP. The results of my research provide critical information about the role of the intensive rice agriculture in the prehistoric Korean diet.

In this study a total of 138 potsherds were collected for the organic geochemical analysis and seven sherds were dated with the luminescence dating. Based on the results of the organic geochemical analyses, each potsherd was assigned to a different food class. Then, these potsherds were ordered in time, based on the results of the luminescence dating and available AMS radiocarbon dating. By doing so, I was able to achieve the primary goal of this research: a re-evaluation of the conventional rice-centered models to better understand the overall pattern of subsistence strategy and assess the weight of rice in it.

1.2. The transition from foraging to farming and the emergence of agriculture

The process of the transition from foraging to farming and the emergence of agriculture are long standing topics of archaeological investigation (Binford, 1968; Childe, 1951; Flannery, 1972, 1976; Redman, 1978). The emergence of agriculture and its role in subsistence is one of the most studied domains in archaeology. The intensification of agriculture and the control over agricultural surpluses have been linked to the origins of the socio-political complexity (Childe, 1951; Earle, 2002; Price, 1995; B. D. Smith, 1989; Welch & Scarry, 1995). A recent collection of papers in Current Anthropology (Vol. 52, 2011) indicates the importance of this topic and diversity of approaches to the transition from foragers to farmers. Current approaches to understanding the subsistence change from foragers to farmers would fall into four categories: (1) population pressure model, (2) climatic fluctuation model, (3) cultural or social model, and (4) evolutionary model.

One of the most well-known approaches is the population pressure model (Binford, 1968; M. N. Cohen, 1977, 2009; Flannery, 1972, 1976). This approach starts with the idea that farming is backbreaking, time consuming, and intensive-labor work. Based on the ethnographic analysis of the Kalahari Desert of South Africa, Binford suggested that even in a marginal area, food collecting was a successful adaptation (1968). Therefore, he argued that human groups would not have become farmers, unless they had had no other choice. Population pressure was therefore suggested as a proper agent for the origin of agriculture: more people required more food. The best solution to the problem, according to Binford, was farming, which provided a higher yield of food per a unit of land. However, at the same time, the intensification of agriculture required more labor to harvest food. M. Cohen (1977, 2009) argued for an intrinsic tendency of growth of human population, which is responsible for the initial spread of the human species out of Africa, and the subsequent colonization of Asia, Europe, and the Americas. Along with this population growth, after about 10,000 BC there was an increase in the use of less desirable resources in many areas. Cohen argued that the only successful way to cope with increasing population and declining resources was agriculture.

The second approach emphasizes climate fluctuation. The role of the rapid climate change in the process of subsistence change is certainly a factor to be considered at various specific points in time (Belfer-Cohen & Goring-Morris, 2011). Bar-Yosef (2011) argued for the rapid climatic fluctuation as the main factor in the origin of the cultivation of various wild plants in East and West Asia. The model is based on the idea that the origin of cultivation was motivated by the vagaries of the climatic fluctuation of the Younger Dryas around 10,000 B.C. within the context of the mosaic ecology which affected the communities that were already sedentary or semi-sedentary. By examining paleoclimatic records with available archaeological phenomena, Bar-Yosef proposed that while the rapid climatic fluctuation served as a trigger of the beginning of cultivation at the end of the Younger Dryas, such changes continued to influence the Holocene period of both East and West Asia.

The third category of approaches focuses more on cultural or social aspects. Cauvin (1994) argued that the important changes associated with the subsistence change from foraging to farming were conceptual as much as, or more than just material (i.e. food production). Specifically, he suggested that farming was led by the emergence of new conceptual ideas such as new cosmology, religious practice, and symbolic behavior. For Cauvin, this transition allowed foragers to view their habitat in a different way and promoted a more active exploitation of their environment. Based on the archaeological phenomena of four cultural areas in China, D. Cohen (2011) argued that the Early Neolithic culture in China, which involved the farming of millet and rice, was invented and spread with a wide range of information exchange and broad social networks rooted in the interactions of Late Paleolithic hunter-gatherer societies (D. J. Cohen, 2003). Recent studies showed that the agricultural origins took place in relatively abundant environment, not in places where little food was available (Price & Gebauer, 1995). This partially supports the idea that the subsistence change from foraging to farming might not be solely explained by the economic aspect.

The last category of approaches is based on evolutionary perspectives. The most well-known study is done by Rindos (1984). He focused on the coevolutionary mechanisms between plants and people during the domestication, incorporating three stages of process: Incidental domestication, specialized domestication, and agricultural domestication. Rindos's explanation for the origin of agriculture can be defined as a neo-Darwinian evolutionary approach. A human is an unconscious agent who selects only for instant benefits. Most models proposed for the origin of agriculture relied on problem-solving abilities of the human to explain this transition: peoples' intent or desire for more sustainable food. But for Rindos, people could not intentionally domesticate plants. However, they did favor those plants that were most useful to them. Man can select, but he could not have known how important the products of their selection would become. Rindos did not address why the agricultural system developed after the end of Pleistocene. He described the question why humans began to establish their coevolutionary relationships with plants as a question without meaning. To Rindos, the explanation for the origins of agriculture will be ecologically specific to each world area where this process took place (Rindos, 1984).

More recent approaches in evolutionary models are based on the evolutionary ecology (Gremillion & Piperno, 2009; Winterhalder & Kennett, 2006, 2009). The evolutionary ecology emerged from an earlier perspective known as 'cultural ecology', which focused on the dynamic relationship between the human society and its environment (Steward, 1972). Evolutionary ecologists have emphasized human ability to reason and optimize their behavior. In this view, the cultural and behavioral change is explained as a form of phenotypic adaptation to changing social and ecological conditions, applying the assumption that organisms are designed by natural selection to respond to their environment in 'fitness-enhancing ways' (Boone & Smith, 1998: 141; Cannon & Broughton, 2010; Winterhalder & Smith, 1992). Archaeologists often assume that hunter-gatherers operate based on the premise of efficiency to obtain sufficient food.

Food is ranked by the energy value it contains; and lower-ranked resources such as seeds are demanded, only as higher-ranked ones become unavailable. In this view, the subsistence change to farming is explained as adding new resources.

Current evolutionary approaches to the subsistence change from foragers to farmers have expanded to sub-disciplines such as the niche construction Theory (Bleed & Matsui, 2010; Crawford, 2011; B. D. Smith, 2007). The niche construction theory emphasizes long-term reciprocal dynamics between humans and their environment, in which modification of their environment helps create the niche they inhabit (Laland & Brown, 2006; Laland, et al., 2001; Odling-Smee et al., 2003). Niche construction by a large number of animal species has been studied in various different regions around the world. Given that so many different animal species manipulate their environments, it is reasonable to assume that humans have been actively managing their environment to varying degrees (B. D. Smith, 2007). Ethnographic studies have documented a growing inventory of the different ways in which human societies actively intervene in their local environments in an effort to shape them more to their liking (B. D. Smith, 2007: 195). In this perspective, agriculture can be one of the acmes of human niche construction.

The rest of the chapters in this monograph will lead us to show which of those models/theories is suitable for explaining the transition from foraging to farming in the central part of the Korean Peninsula by incorporating the innovative analytical methods: the organic geochemical analysis and Luminescence dating.

1.3. What do we know so far about transitions to agriculture?

Some of the studies that I have mentioned above show that in some parts of the world, farming spread rapidly and patchily from one place to another. However, other studies indicate that it spread very slowly in other areas; in some places people did not become farmers for up to a millennium after their initial contact with agriculture, or never became farmers at all. Sometimes these areas are environmentally segregated (e.g. Alps or Pyrenees), but can be also defined by social factors (Robb, 2013). If we think of places that show any evidence of farming (for example, Europe, which is the most thoroughly studied region in relation to the emergence of agriculture and spread of farming), there are several underlying characteristics these areas have in common, which will be discussed from now on (Robb, 2013; Whittle & Cummings, 2007).

1.3.1. Migrations of farmers

Though it is highly varied in form, it is true that there were actual movements of farmer/farmers from one place to another. However, at the same time, there is no real evidence for a massive migration in terms of a single big wave of movement which covered large landscape. In fact, most archaeologically traceable human movements are 'opportunistic leap-frog' (Boland, 1990; Robb, 2013: 658) migrations. These movements seem to involve small groups of people with no typical single origin, resulting in a complicated form of migration without homeland.

1.3.2. Genetic studies

Unfortunately, unlike the initial optimistic views (Cavalli-Sforza et al., 1994), the results of genetic studies are quite ambiguous and inconclusive. Though several researches showed that there is genetic discontinuity between hunter-gatherers and early farmers, and between hunter-gatherers and the modern population in some places (Malmström et al., 2009; Rowley-Conwy, 2009), other studies suggest that both incoming and indigenous peoples contributed to the gene pool of the modern population(Bramanti et al., 2009; M. Richards, 2003).

1.3.3. First contact

In many cases, when there is contact between foragers and farmers, the former often adopt new subsistence strategies (such as farming) little by little for their own sociopolitical purposes (Robb, 2013). This is somewhat different from the traditional view that new economic practices (based on farming and animal domestication) with innovative technologies (notably, pottery and new types/forms of stone tools) rapidly spread into the foraging context as a 'package', completely transforming the society to a fully farming community (Childe, 1951).

Summing up, if there is any conclusion that archaeologists can reach, it would be that the transition from foragers to farmers and spread of farming occurred in a 'mosaic way' (Robb, 2013: 659). This means the transitions occurred around the world in various and diverse ways. This diversity motivates us to investigate the specific manifestations of this transition in different parts of the world and better understand the different ways that people made this profound transformation.

1.4. Cases in East Asia

Since the main study area of my research is the central part of the Korean Peninsula, it is worth to examine the transition from foraging to farming and the emergence of agriculture in its neighbors, namely China and Japan, to provide some regional context. Numerous archaeological and historical studies showed the similarities between the material cultures from those three regions (Nishitani, 2014; Noh, 2003; Y. S. Seo, 1981). Geographically, China is located in the west of the Korean Peninsula and its northeastern boundary is bordering the north of North Korea (Figure 1.1). The Japanese archipelago extends from northeast to southwest along the east side of the Korean Peninsula (Figure 1.1).

1.4.1. China

The critical time period related to the transition from foraging to farming and the emergence of agriculture in China is that between 12,500 and 9,000 cal BP when hunter-gatherers in four distinct geographical 10 regions (Northeast China, the North China plains, and the Middle and Lower Yangtze River regions) established the first sedentary villages (D. J. Cohen, 2011). Recent debates have been focusing on the timing and the speed of this transition (Crawford, 2009; Fuller et al., 2009, 2010; Liu et al., 2007; Zhao, 2011), investigating how harvested crops (e.g. rice and millet) were incorporated into changing the mode of subsistence over a 3,000-year period (D. J. Cohen, 2011: 29). Unfortunately, it is still not clear where agriculture first started. The lower Yangtze region, where rice agriculture begins, was assumed to have the earliest evidence of plant domestication. However, new data showed that the northern plains of China have the independent tradition of early millet farming (Bar-Yosef, 2011; Barton et al., 2009; Bettinger et al., 2007; Lu, 2006; Shelach, 2000). Although the fundamental reliance on the harvested crops such as rice and millet was once thought to be a major part of the change, in recent years it has become clear that this is not the case. The advent of plant domestication and the subsequent agriculture was a slow process in a number of small steps, region to region (D. J. Cohen, 2011; cf. Robb, 2013).

1.4.2. Japanese Archipelago

Traditionally, in Japan, the strict dichotomy between Jomon Neolithic hunter-gatherers and Yayoi Bronze Age farmers persisted among the archaeologists. This trend began in the 1980s when the concept of affluent foragers was considered to provide a suitable explanation of Jomon economy (Aikens et al., 1986; Koyama et al., 1981). However, this widely accepted view that the Jomon people sustained their hunter-gathering life for several thousands of years in a "naturally rich environment (Crawford, 2011: S336)" was criticized by other scholars, for it oversimplifies the Jomon subsistence (Crawford, 2008; Kobayashi et al., 2004; M. Nishida, 1983). A recent study from Obata and his colleague (2007) showed solid evidence of plant domestication during the Middle Jomon period in the Kyushu area.

To most Japanese archaeologists, farming is considered as irrigated rice agriculture during the Yayoi period. However, according to Crawford (2011) the Yayoi agricultural system was not solely based on rice. A wide range of plants including millet, barley, wheat, and leguminous was also a significant component of the Yayoi agricultural economy.

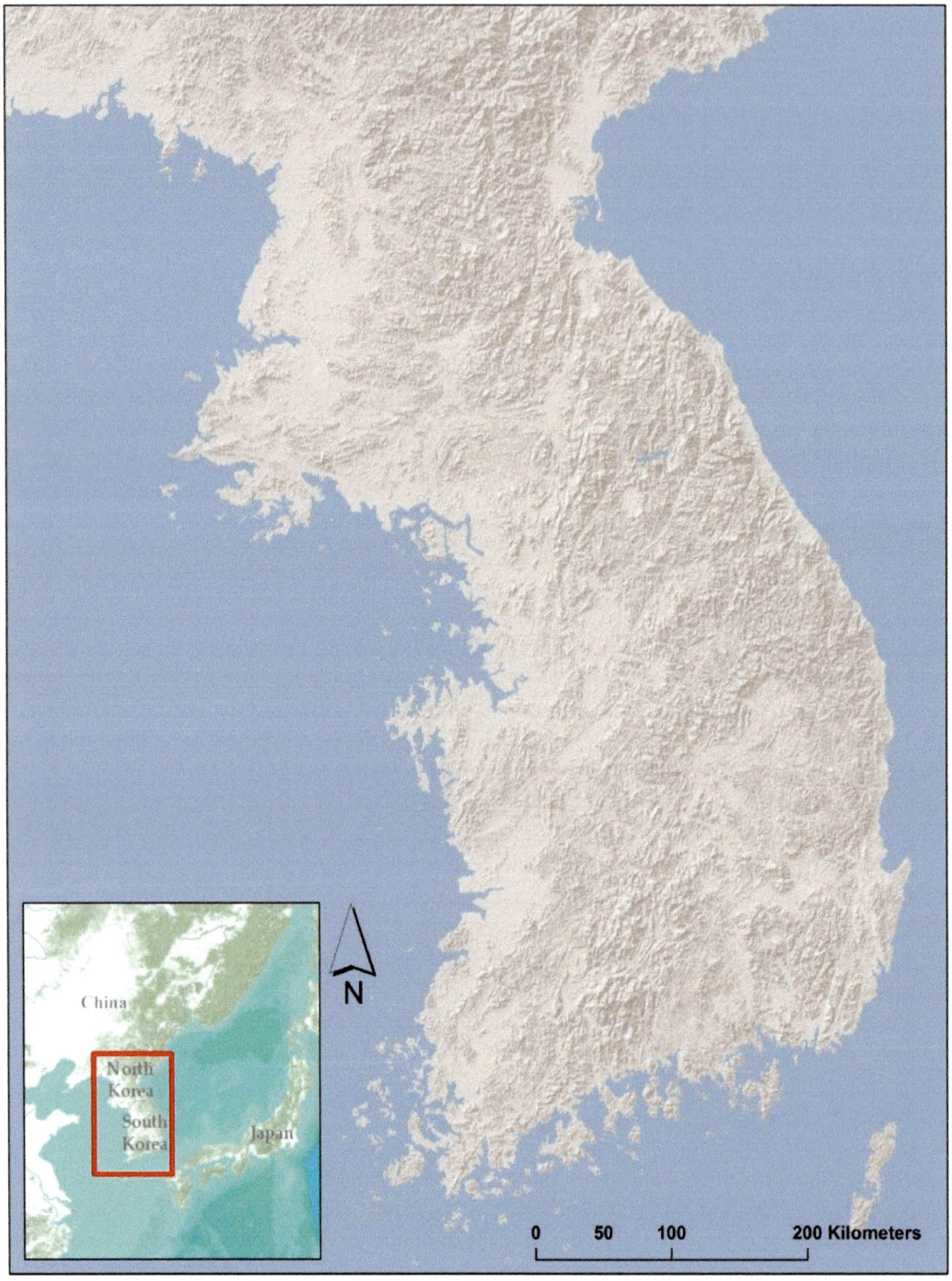

Figure 1.1: The map of the Korean Peninsula and its vicinity

2. Background and Central Hypothesis

2.1. Archaeology in Korea - Its brief history and social context

Although there are some regional dialects, in regard of its culture and language, Korea includes no recognized minorities. Therefore, traditionally, the Korean prehistory is frequently formulated in Korea with reference to ethnicity, perceiving the elucidation of the formation of the Korean people to be the chief purpose of archaeology. M. K. Kim (2008) made an interesting observation related to the history of Korean archaeology and its social context.

In the twentieth century, the Korean Peninsula underwent a series of dramatic political upheavals. This political fluctuation began with the Japanese annexation of the country in 1910. The liberation of the Korean Peninsula in 1945 after the end of the World War II was followed by the Korean War (1950-1953) and the subsequent establishment of two competing states: the Republic of Korea (South Korea) and the Democratic People's Republic of Korea (North Korea). This political context established a particular and unique social milieu, which critically influenced archaeological practices. The modern practices of archaeology in Korea were first conducted by Japanese archaeologists such as Tadashi Sekino, Ryuzo Torii, and Ryu Imanishi during the colonial period. Archaeological remains, which are inherently subject to a variety of interpretations, were easily exploited to justify the Japanese colonization of Korea (M. K. Kim, 2008). Through this, Japanese archaeologists tried to claim that the Korean people were characterized by "a lack of independence" and "a servile attitude towards bigger nations." Though it seems that this is a typical example of colonialist archaeology of Trigger (1984, 2008), there is a huge difference between the one and the other. The colonizers were Japanese, not Europeans. Though one might argue this is unimportant, in fact, it is. While European colonizers did not have any cultural or historical similarities with Native Americans, Japan and Korea have actively been interacting to each other since the Late Neolithic Age. For this reason, the archaeological phenomena of Korea and Japan are quite similar. Therefore, Japanese archaeologists who practiced archaeology in Korea argued that all prehistoric/historic material cultures were handed down from the Japanese isles to the Korean Peninsula. The primary character of the colonialist archaeology defined by Trigger is denigrating native peoples by presenting the primitive aspects of their archaeological phenomena. However, in this case, the Japanese justified their colonization by emphasizing the overall similarities and excellence of the prehistoric/historic material cultures of Korea and Japan.

As in many post-colonial nations, the Korean archaeology after the liberation from the Japanese colonization has took a central role in refashioning the national identity and restoring the national pride (M. K. Kim, 2008). Especially in South Korea, archaeological phenomena have been being interpreted as evidences of migration and cultural diffusion throughout the Eurasian continent. Highlighting a harmonious blending of different cultural traits and emphasizing cultural interactions over a vast region may appear to contradict nationalism which assumes the ethnic superiority. However, it should be noted that such interpretations describe the ancient Koreans as a people with a grandiose geographical scope whose life was not confined to a small peninsula. The interpretations of the archaeological phenomena in Korea often intentionally aim at suggesting the creativity and superiority of the Korean people. Based on this, some archaeologists have recognized nationalism in the Korean archaeology and have described the current Korean archaeology as nationalist archaeology (M. K. Kim, 2008; Trigger, 2008).

However, in the middle of the 1990's, archaeology in Korea started to make various voices heard. The second generation Korean archaeologists who were educated in the United Kingdom and the United States as 'graduate students' began to conduct their own researches in Korea. Though they were highly influenced by the nationalism of the Korean archaeology from the first generation archaeologists, they also learned major theoretical frameworks and empirical methodologies from decent universities in US and UK. Currently, these scholars, on one hand, are trying to avoid an extreme nationalism, and on the other, they are also concerned about the imperialist aspect of their knowledge originated from UK and US.

The studies on the emergence of agriculture in Korea went through a similar trajectory. As I mentioned in Chapter 1, the Koreans' attachment to rice is remarkable. Regardless of their economic status, way of life, or ideological inclination, steamed rice was and is an essential dish throughout the nation. Within this social context, over the last 50 years, one of the main tasks of the Korean archaeology has been finding the earliest evidence of rice agriculture (or even just rice). Of course, the intention behind this was to promote the nation's identity as rice-eating Koreans. The most extreme case was from Soro-Ri, a small town in the central part of the Korea Peninsula (M. K. Kim, 2008). The Soro-Ri site (the site was named after the town, which is common in the Korean archaeology) is an Upper Paleolithic site with an estimated date of approximately 30,000–20,000 years ago (Han & Son, 2000). It was considered as nothing but just an ordinary Paleolithic site. However, everything was changed when some of the archaeologists who participated the excavation claimed that they had found the oldest evidence of rice in the world (Y. J. Lee & Woo, 2002). The claimed evidence was the husks of rice grains, and their AMS radiocarbon date (12,500±2000 BP, uncalibrated) preceded any other directly dated rice remains found in the world (cf. Higham & Lu, 1998). The site rapidly became famous outside of the academia. Major newspapers and television news treated this discovery and linked this with the Korean people's "spirit" or "blood" (S. H. Kim, 2004; Kwak, 2005). These statements obviously show the nature of nationalist archaeology. However, most recent archaeological studies on agriculture conducted by second generation archaeologists try to avoid an extreme nationalism (cf. M. K. Kim, 2015; G. A. Lee, 2011). I consider myself as a third generation archaeologist in Korea, for I learned archaeology as an undergraduate student from second generation archaeologists, and by studying archaeology in US as a graduate student. In this perspective, I will try to avoid both the nationalist and imperialist aspects in my interpretation on the results of this study.

2.2. Chulmun foragers and Mumun farmers - where everything started

This monograph investigates the process of transition from foraging to farming and the role of agriculture as a subsistence strategy during this transition in the central part of the prehistoric Korean Peninsula (Figure 2.1a). The period in question has been called the Mumun pottery period (3390-2290 calibrated years (cal.) B.P., cf. Bale, 2012). The traditional periodization scheme of the prehistoric Korea is based on the decorative attributes consistently found on the potteries that existed over specific time periods: 9950-3390 B.P. is the Chulmun (Chŭlmun, Jeulmun, or 'comb-pattern') Pottery period and 3390-2290 cal. B.P. is the Mumun (or 'undecorated') Pottery period (Bale, 2012; Norton, 2007). Sometimes the former and the latter are respectively regarded as the Neolithic, and the Bronze Age of Korea (Ahn, 2004; Norton, 2007). The beginning of the Mumun period has an important role in the Korean archaeology, for it has been linked with the beginning of the agricultural society. The Mumun period, named after its representative pat ternless feature of pottery, is known for the intensive rice farming, instead of hunting and gathering of the Chulmun period. Also, with this economical evolution, the society became more complex and a social hierarchy emerged. 'Mumun', term meaning 'undecorated', is the most common feature of the pottery in this period. Ahn Jae-ho devised this influential 'Chulmun-Mumun' periodization based on diagnostic changes in pottery decoration, pit-house architecture, interior pit-house features, and stone tool types (J. H. Ahn, 1991, 2000, 2001). Ahn's chronology assumes that changes in pottery decorative attributes and plan-shapes of pit-houses are time-sensitive. According to him, the Mumun periodization scheme has the following internal stages: Incipient, Early, Middle, and Late.

Korean archaeologists have been focusing on the differences between the overall archaeological assemblages of the Chulmun and Mumun periods. Now, I will briefly examine the different aspects of the archaeological assemblages from the two periods.

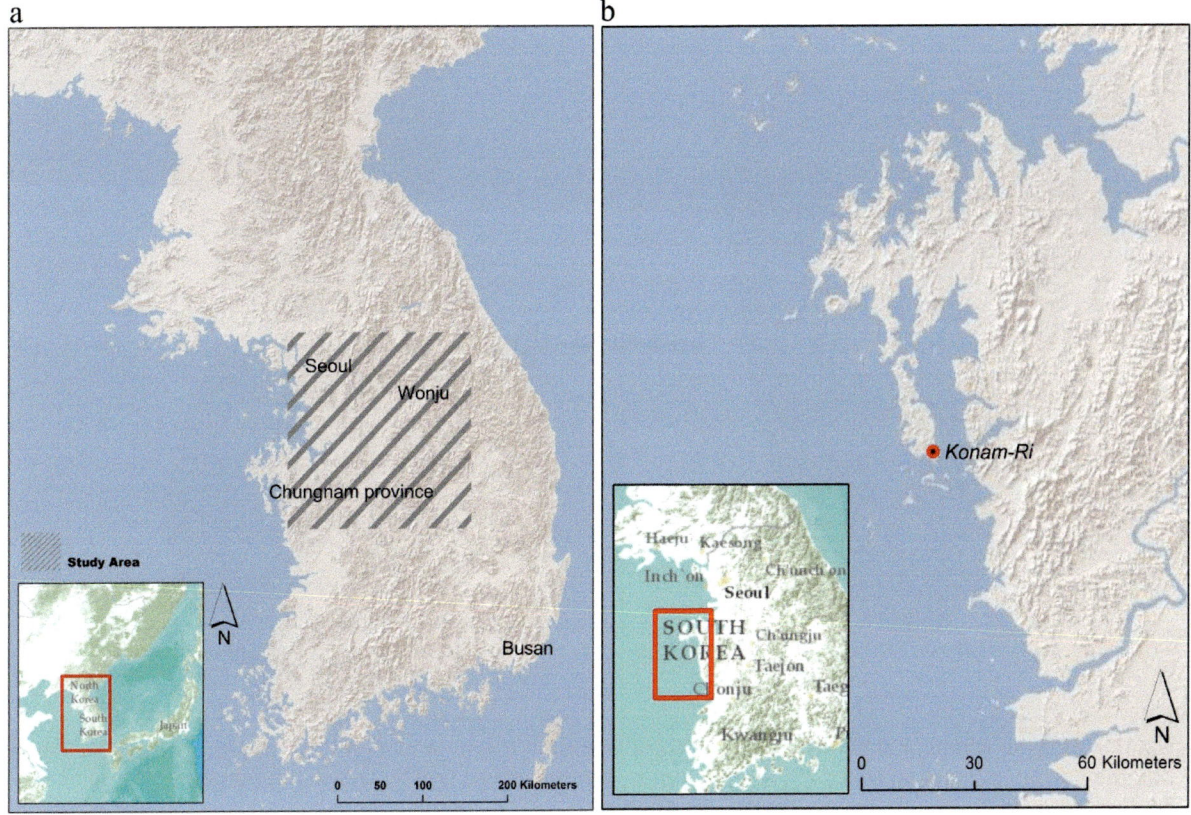

Figure 2.1: (a) The indication of the central part of the Korean Peninsula (b) The location of the Konam-Ri shell midden

To begin with, in the case of pottery, the fundamental characteristics of the Chulmun period pottery are the comb-shape pattern and the pointed bottom, which show some variations as the phases go by (Figure 2.2a). Some pieces of the Chulmun period pottery from the Gangwon province (Figure 2.7) have the flat bottom, but this shape is considered as an exception to the general form of the Chulmun period pottery. On the other hand, all the Mumun period pottery have the flat bottom; the major part of their body does not have any pattern. Some patterns still exist, but are confined to the extreme upper body (Figure 2.2b). During the incipient stage of Mumun, potteries had a pinched clay strip attached to the outside of the rim and body (Cheon, 2005) (Figure 2.7; 2.4a). The Early Mumun pottery have both rim-punctuations and lip-scoring. This combination of attributes is sometimes referred to as Yeoksam-Dong-style (Figure 2.7; 2.4c) after the site where they were first uncovered (B. G. Lee, 1974). Another pottery style of the Early Mumun, Garak-Dong-style (Figure 2.7; 2.4b), is named after a site in Seoul, but settlements with this pottery tradition are found clustered in the tributary valleys of the Geum-gang River (B. G. Lee, 1974). Garak-dong style deep-bowls have appliqué rims (or double rim) with short slanted lines that are incised just below where the rim attaches to the body. The last type of the Early Mumun potteries is the Heunam-Ri-style pottery (Figure 2.7; Figure 2.4d), which is a combination of Yeoksam-Dong and Garak-Dong styles (J. H. Ahn, 2000: 49; J. S. Kim, 2001; S. H. Lee, 2005). From the Middle Mumun period, potteries become completely undecorated. The most dominant one is the Songguk-Ri-style pottery (Figure 2.7; Figure 2.4e) which have elongated and curved shapes with everted rims in comparison with the Early Middle Mumun pottery (Norton, 2007).

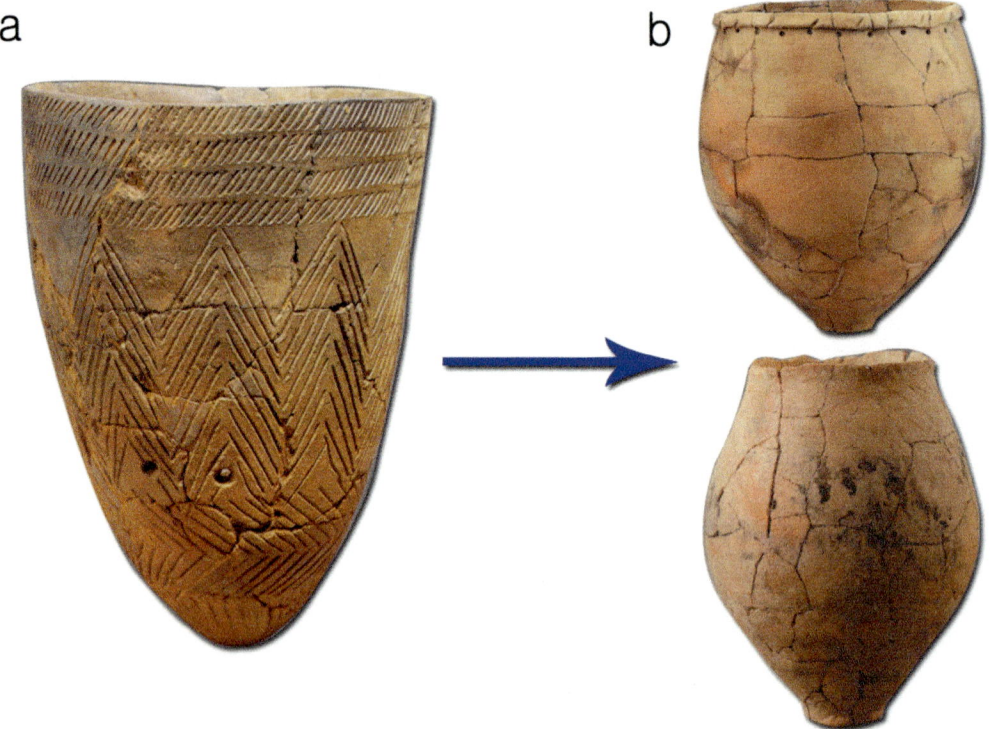

Figure 2.2: The Chulmun and the Mumun period potteries (a): a Chulmun pot with the comb-shape pattern and pointed bottom (b): two pieces of the Mumun pottery with patterns mostly on the rim (upperright; Heunam-ri-style) and with no pattern (down-right; Songguk-Ri-style)
(modified from Yoon & Bae, 2010)

The manufacturing technique of stone tools too shows many discrepancies between the two periods. Though polished stone tools started to be used in the Chulmun period, their qualities and the skill of their production are relatively poorer than those of the Mumun period (Figure 2.3a). The stone tools of the Mumun period including the polished stone arrowhead and dagger, which were excavated in the central part of the Korean Peninsula, are very elaborate and exquisite (Figure 2.3b). Also, from the middle of the Mumun period, we begin to observe bronze ware.

The form of habitations also changes. The Chulmun period's houses have generally a round shape, but this shape was transferred into a rectangular style longhouse in the Mumun period (Figure 2.5). Inside the longhouse, we can observe a row of 3 or 4 hearths for warming/cooking, which are not seen in that of the Chulmun period. In a few words, the Chulmun period's pottery with the pointed bottom and comb shape pattern, and its polished stone tools and round-shape habitation were changed into the patternless flat-bottom pottery, elaborate polished stone tools and rectangular-shape habitation.

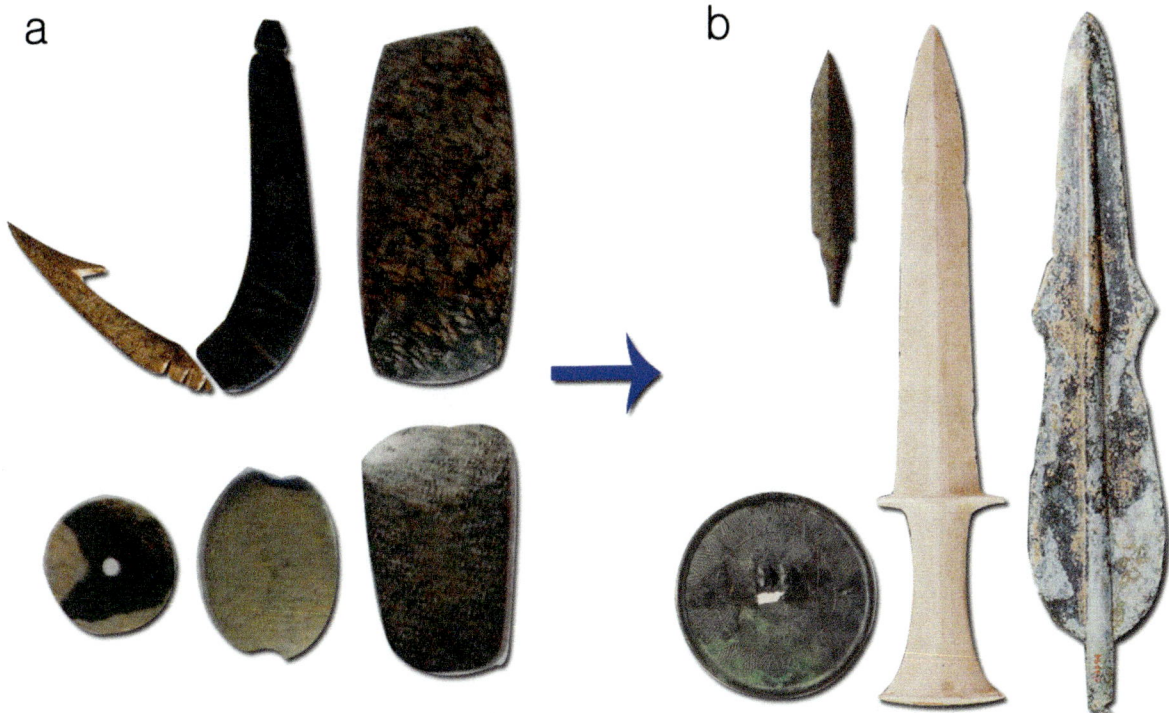

Figure 2.3: The Chulmun and Mumun period tools (modified from Yoon & Bae, 2010)

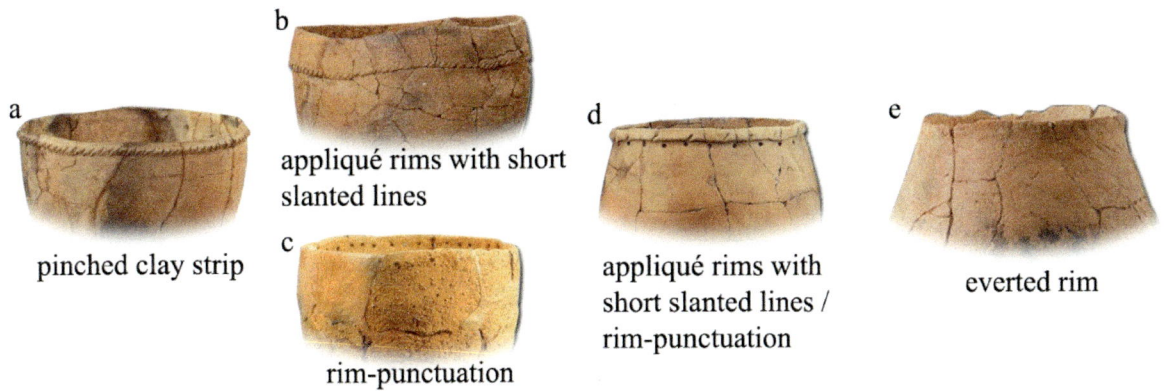

Figure 2.4: The patterns on the Mumun pottery (a): pinched strip (Cheon, 2005) (b): Garak-Dong style (B. G. Lee, 1974) (c): Yeoksam-Dong style (B. G. Lee, 1974) (d): Heunam-Ri style (J. H. Ahn, 2000) (e): Songguk-Ri style (Norton, 2007)

Together with these differences in characteristics of the archaeological assemblages of the two periods, Korean archaeologists assume that the most distinctive difference between the two periods consists in their subsistence strategies. Agriculture brought a great change into human life. Engaging in farming, human beings settled down for the first time. In the Korean Peninsula, it is argued that in the Mumun period agriculture became the main means of living due to rice. Clear evidence including stone sickles (Figure 2.6a), "semi-lunar shaped" stone knives (Figure 2.6b), as well as dry field (Figure 2.6c) and irrigated rice paddies (Figure 2.6d) shows that the full-dress farming was practiced in this region around the beginning of the Mumun Period (G. A. Lee, 2003, 2011; Yoon & Bae, 2010). Korean archaeologists think that agriculture was introduced in the Chulmun period's late phase and rice agriculture spread widely in the Mumun period's early phase to be the principal subsisting way in the Mumun period's middle phase. They think that though agriculture was introduced during the Chulmun period, the main subsistence in this period was confined to hunting, fishing, and gathering. Normally, the start of rice agriculture is treated as being very important; and the site that gave initially grains of rice, burned or not, is thought to have a critical meaning. However, what matters is not the start of rice agriculture, but its general practice. Korea is an agrarian country even nowadays, and rice is still the staple food of the Korean people. Therefore, it is essential to know when the ancients of the Korean Peninsula started to eat rice as staple food.

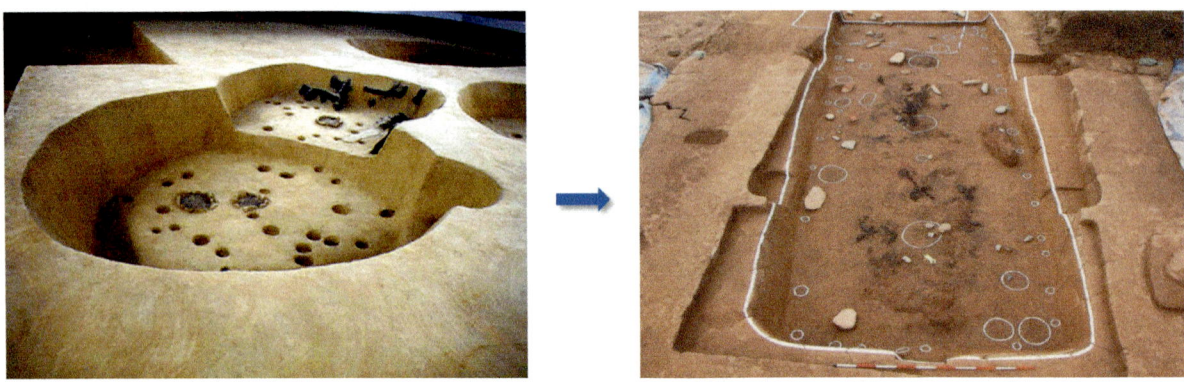

Figure 2.5: The Chulmun and Mumun period habitations (modified from Yoon & Bae, 2010)

2.3. Current views on the transition from foragers to farmers and development of rice agriculture in the Korean Peninsula

According to the G. Lee (2011: S322), the transition from foragers to farmers in the Korean Peninsula has been approached by assuming a strict discrepancy between Chulmun hunter-gatherers and Mumun full-dress rice farmers (J. H. Ahn, 2000; B. C. Kim, 2006a). The transition was linked with multiple migration events coinciding with the climate change (J. S. Kim, 2003, 2006), or assumed to be driven by the population growth (Norton, 2000, 2007), or regarded as consequence of a risk reduction strategy (J. J. Lee, 2001).

Until recently, analyses of marine resources from coastal shell middens have been the primary data source for investigating patterns of subsistence in Korea (cf. J. J. Lee, 2001, 2006; Norton, 2000, 2007). For example, J. J. Lee (2001) argued that people used farming as a risk-reduction strategy against the declining sea level on the east and south coasts, as the ratio between the population and marine resources became imbalanced after 4,000 BP. By comparing the results of the analyses of marine resources from the shell middens of the west, east, and south coasts, J. J. Lee argued that farming emerged to overcome the loss of marine resources along the east and south coasts. Similarly, Norton (2000) stressed the population growth as one of the key factors for the adoption of rice farming along coastal settings. He examined the remains of marine resources from the Konam-Ri shell midden (Figure 2.1b), located on the west coast of the Korean Peninsula. Based on the results of this examination, he suggested that the differential processing of big fish might be an evidence of residential stability. Residential stability, he argued, led to the increased population throughout the hunter-gathering stage. This population increase, and the associated increased human predation, caused a decrease in the size of fish and other favored taxa, and subsequently pushed the hunter-gatherers to adopt rice farming (Norton, 2000).

J. Kim (2003, 2006) suggests a combination of environmental fluctuation and subsequent human migrations from northern latitudes as a major factor of the agricultural transition in the central part of the Korean Peninsula. Based on paleoclimate data for the early Holocene East Asia, he argued that because of the cooling climate and decreasing temperature around 4,000–3,000 BP, the farmers in the Jilin-Duman regions along the current border with China might have migrated to the central part of the Korean Peninsula, which was better suited for farming (cf. G. A. Lee, 2011: S322). He presented a sudden change in household pattern and the presence of finely ground stone daggers around the central part of the Korean Peninsula as evidences of these migrations. In addition, Kim assumes that the mobility of indigenous hunter-gatherers was constrained when immigrant rice farmers blocked their way to resource patches. The inaccessibility of foraging areas enhanced the transition of hunter-gatherers to farmers (B. C. Kim, 2006b).

Lastly, B. Kim (2005, 2006a, 2006b) focused on the emergence of a complex society associated with an intensive rice agriculture around 2,600BP. By correlating regional scale survey data from the south-eastern Chungnam province (Figure 2.7) with its soil productivity for rice agriculture based on a site catchment analysis of the region, Kim argued that the emergence of a social hierarchy and the subsequent social complexity were driven by the rapid spread of the intensive rice agriculture into foraging contexts. He asserted that this rapid transition is exemplified by the sudden presence of harvesting tools of ground stone.

There are two underlying key ideas that these studies have in common, but both are problematic. The first two studies assume that shell middens can represent the general process of subsistence change from foragers to farmers in the central part of the Korean Peninsula. Since a peninsula, consequently the Korean Peninsula is a part of a continent, the data from the coastal shell middens cannot represent the subsistence of the inland, which includes considerably large habitation sites. Next, all the four studies assume rice to be a dominant subsistence resource since 3,400 BP, without considering the possibility of the utilization of a more wider range of resources for subsistence.

According to paleoethnobotanical evidence from the southern part of the Korean Peninsula, which includes the Daundong site in Ulsan and several localities within the context of the Nam River in Jinju (Oun I, Okbang 1,2,4,6 and 9, Sangchon B), the diet of the ancient farmers of the region included various resources such as millet, soybean, and azuki between 3400 and 2,600 BP. (Crawford & Lee, 2003; G. A. Lee, 2003, 2011) (Figure 2.7). I assume the subsistence pattern might be similar in the central part of the Korean Peninsula during this period, though we lack, for the moment, clear paleobotanical evidences to test this assumption. Therefore, the re-evaluation of those rice-centered models is required, and the general chronology of subsistence during this period has to be established.

Figure 2.6: The evidence of the full-dress farming in the central part of the Korean Peninsula: (a) stone sickles, (b) semi-lunar shaped knives, (c) excavated dry field, and (d) irrigated rice paddy
(all modified from Yoon & Bae, 2010)

2.4. The central hypothesis of this study

My central hypothesis in this study is that there was utilization of a wider range of (wild) animal and plant resources along with rice among ancient farmers in the central part of the Korean Peninsula between 3,400 and 2,000 BP. Studies have shown that in some cases, the initial domestication of crops and subsequent agriculture appeared as a part of the complex foraging economy in an affluent environment (Price & Gebauer, 1995; Price & Bar-Yosef, 2011) and hunting, gathering and fishing persisted well after farming was introduced (Borić, 2002; Craig et al., 2011; Galili, et al., 2003; Milner et al., 2004). In the Yangtze River Valley in China, for example, as well as in the Sub-Saharan Africa and the eastern North America, evidences of very early domestication come from settlements situated in zones with very rich resources which are associated with river valleys, and in none of these areas does domestication appear to have developed within a context of population growth forcing humans into marginal environmental zones (B. D. Smith, 2007). New strategies such as agriculture were initiated by relatively complex hunter-gatherers in circumstances where risk is affordable. Then why did these foragers invest their efforts in agriculture when there was no immediate risk? The key idea for the reply to this question is that an increased sedentism was a "pre-requisite" for the advent of agricultural societies, for complex hunter-gatherers are characterized by a relatively large population and sedentism (G. A. Lee, 2011; Price & Gebauer, 1995: 8). Recent case studies in the eastern North America by Smith (B. D. Smith, 1995, 2007, 2011) are good examples. Smith argued that many of our present domesticated plants originated from the weeds growing in open habitats created by rivers (e.g. floodplain), and they were easily adapted to open areas in the habitats disturbed by human sedentary settlements. Those weeds that invaded open areas in human settlements eventually became domesticated in conformity with the natural outcome of the selective relationship between people and plants within a stress-free environment (B. D. Smith, 2007, 2011). Even the Jomon Japan, the period that is traditionally considered as giving an "affluent" hunter-gathering context based on sedentism, showed clear evidences of plant domestication (Obata et al., 2007). Recently, Crawford (Crawford, 2011) stressed that the orthodox view that the Jomon sustained hunting and gathering for millennia in a naturally rich environment is oversimplification if not correct.

This situation could have existed in the prehistoric Korea. We have solid evidences of a long-term, permanent occupation of the peninsula by complex hunter-gatherers at various places since around 6,000 BP. At the Amsa-Dong Site (Figure 2.7) in the south-east Seoul, at least 12 houses, a significant amount of pottery and different types of ground stone tools such as arrow points, spear points and sickles, were excavated (Im, 1985). Considering that the site was not fully excavated, and based on the scale of the houses as well as the diversity of ground stone artifacts, we can easily assume that this provides clear evidences for sedentism. The house structures and seasonality of the faunal assemblages at the Tongsam-Dong site (Figure 2.7) in the southern part of the Korean Peninsula indicate that people lived there year-round on a permanent basis (G. A. Lee, 2011; J. J. Lee, 2001). We have pollen data from 5,500 BP to 2,600 BP showing that there were specific subsistence solutions which include distinctive combinations of wild (e.g. acorn (*Quercusacutissima*Carr.), Manchurian walnut (*Juglans* spp.)), possibly managed (e.g. chenopod (*Chenopodium* sp.), panicoid grass (*Paniceae*)), and domesticated (e.g. foxtail (*Setariaitalica* ssp. *italica*) and broomcorn millet (*Panicummiliaceum*), possibly soybean (*Glycine max*), azuki (*Vignaaugularis*) and beefsteak plant (*Perillafrutescens* (L.) Britt)) plants (G. A. Lee, 2011: S326). On the other hand, though we lack the evidence of faunal remains due to the high acidity of sediment in the Korean Peninsula, it is still possible that hunting and fishing may have persisted along with farming after its introduction (cf. Craig et al., 2011; Milner et al., 2004).

In this regard, the prevailing rice-centered models, which assume rice to be the most dominant subsistence resource since 3,400 BP., are misleading. What is overlooked in the subsistence studies of the prehistoric Korea is the distinction between the first adoption of crops and the later development of the intensive agriculture (G. A. Lee, 2011). The migrants (cf. J. S. Kim, 2006) probably needed time to adjust themselves to the local environments, especially for rice agriculture, which required complicated irrigation techniques. As G. Lee (2011) noted, rice may have played a minor subsistence role at this time, and it may not have served as a driving factor of the emergence of social complexity.

2.5. The Chulmun and Mumun periods: Essentialism vs. Materialism

Before I move on to the next chapter, I would like to further examine prevailing concepts of the Chulmun and Mumun periods in the Korean archaeology, focusing especially on potteries. As I indicated in chapter two, the potteries of the two periods have several key physical traits which have lead Korean archaeologists to consider the discrepancy between those of one period and those of the other. For example, the fundamental characteristics of the Chulmun period potteries are the comb-shape pattern and pointed bottom (Figure 2.2a). On the other hand, all the Mumun period potteries have the flat bottom, and the major part of their body does not have any pattern. In some cases some patterns still exist, but are confined to the extreme upper part of the body.

At a first glance it sounds quite reasonable to divide the two periods, especially when we compare the Chulmun potteries showing the extensive and intensive comb-shape pattern, and the mostly unpatterned Mumun potteries (Figure 2.2). However, if we examine closer, there are some variations in characteristics which have been somewhat neglected. Until now, probably the most well-known Chulmun period pottery is the one from Amsa-Dong (Figure 2.7; 2.2a). The entire body of a pot is decorated with comb-shape patterns which can be divided into three different parts (Figure 2.7a). The pattern in each part has a different length and a different angle which makes a distinctive characteristic of the pottery. This was the earliest form of the Chulmun potteries in the central part of the Korean Peninsula, and appeared around 6000 BP. However, the pattern on the Chulmun potteries gradually changes as time goes by. Figure 2.8b presents a Chulmun period comb-shape patterned pot found at an upper layer of the Amsa-Dong site.

Interestingly enough, the patterns on its bottom part are gone and those of its middle part became less distinctive. If we see the Late Chulmun period potteries excavated from the Amsa-Dong site, Seoul city and the Sammok island site, Incheon city (Figure 2.7c; 2.7d), we can verify the patterns on their middle part also vanished away and only those on their top part remain. These latest Chulmun potteries from Sammok island and Amsa-Dong give an important clue to the relationship between the Chulmun and Mumun potteries: the pattern in this stage of the Chulmun potteries exists only on their top and rim, and the most distinctive characteristic of the Incipient/Early Mumun potteries is various patterns on their rim (Figure 2.4). These rim-based decorations make the two potteries seem similar. Some of the Chulmun potteries have even the rim-punctuation (Figure 2.7e) which is commonly observed on the Early Mumun potteries (Figure 2.4c; 2.4d). These similarities in pattern between the late Chulmun, and the Incipient/Early Mumun potteries suggest a close connection between them. All this contrasts with the current dominant idea which assumes the discrepancy between the Chulmun and Mumun periods (B. C. Kim, 2006a, 2006b; J. S. Kim, 2003).

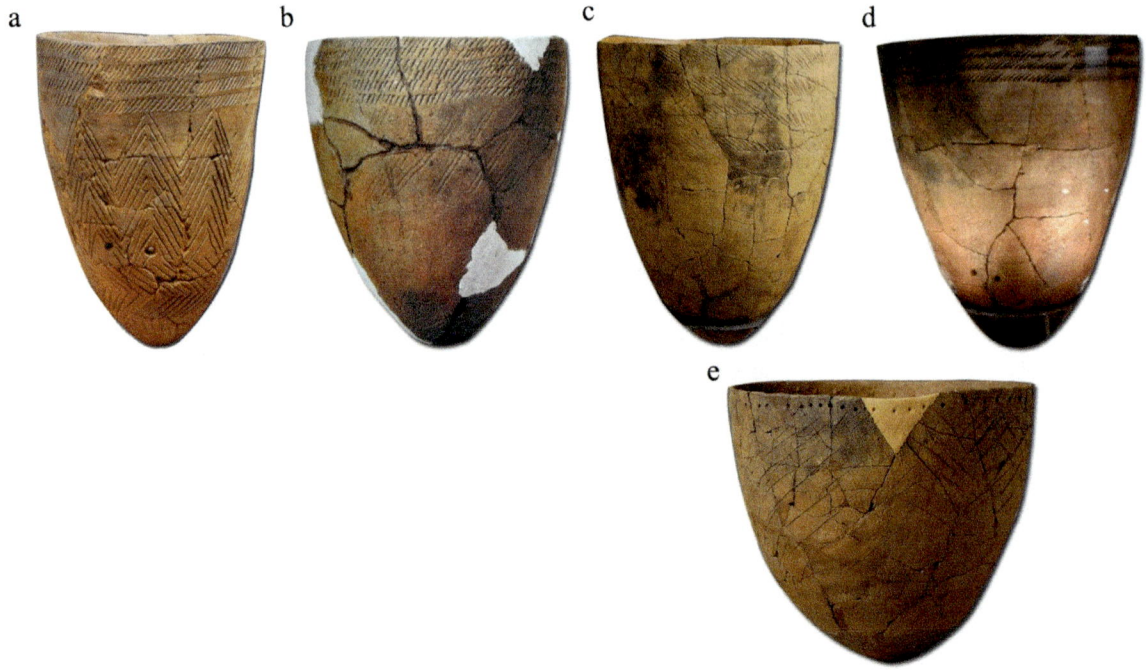

Figure 2.7: Variation in pattern on the Chulmun potteries (a): Amsa-Dong (b): Amsa-Dong (c): Sammok island (d): Amsa-Dong (e): Yongyou island

If we consider the relationship between the Chulmun and Mumun potteries in light of the two different perspectives of essentialism and materialism, the picture is clearer. Essentialism is a philosophical stance which supposes the existence of a specific entity which can be both identifiable and distinguishable. If things share actual and fixed characteristics, these essential traits can be used to distinguish group A from another group B, and constitute the essence of each of the two groups. The most prominent point of this 'essence' is its unchanging permanency. In contrast to essentialism, materialism holds that phenomena cannot exist as fixed entities, because they are always in the process of becoming something else. In materialists' view, things are in a state of flow: no two things can ever be put into the same category, because even similar things are just at similar points in the process of becoming others. In the discipline of biology, the two philosophical stances were identified by Ernst Mayr (1959) in defining the concepts of biological species: 'typological' thinking versus 'population' thinking (Marwick, 2008, Figure 2.8). A key point in differentiating essentialism from materialism is not that the former treats difference and the latter change, but that the one treats only difference while the other treats both difference and change (O'Brien & Lyman, 1998: 29).

For many years, archaeologists in Korea have been studying sites and artifacts within the framework of ontological essentialism by the name of 'typology' or 'classification'. Archaeologists often use the terms classification and typology interchangeably, but a distinction must be made between them. A classification is any set of formal categories into which a particular field of data is partitioned, while a typology is a particular type of rigorous classification, in which a field of data is divided up into the categories that are all defined according to the same set of criteria, and that are mutually exclusive (W. Y. Adams, 2001; W. Y. Adams & Adams, 2007). Therefore, to be precise, it is not classification but typology that many archaeologists have been employing for studying their sites and artifacts. They grouped artifacts according to some characteristics for demonstration of cultural traits and cultural changes. Objects are split into categories —in other word, 'types' — according to their perceived similarities, and change is viewed as transition from one type to another. This means that as long as the objects are in the same category, they are closer to each other than any other objects in different categories (even if one category comes right after another in terms of time). By doing so, archaeologists have been creating units for their interpretation of archaeological phenomena. However, there is nothing "inherent" in the units they use that makes them real (O'Brien & Lyman, 1998: 30). Artifacts may share certain traits in common which make us put them in the same category, but there is no reason to think each category is genuine.

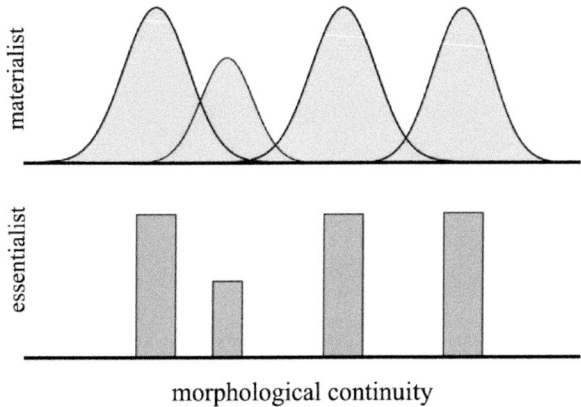

Figure 2.8: Schematic representation of the essentialist ('typological' thinking) and materialist ('population' thinking) approaches (modified from Marwick, 2008: 108)

A similar trend has been prevailing in the Korean archaeology, especially when we consider the transition from Chulmun to Mumun. Whether they recognized or not, Korean archaeologists created two units, named them "Chulmun pottery" and "Mumun pottery", and regarded the two as different entities. In addition to that, they expanded this concept to the entire archaeological phenomena of the two periods. Since change is viewed as transition from one separate entity (Chulmun) to another (Mumun), everything in the Chulmun period has to be drastically different from everything of the Mumun period, including pottery, house pits, stone artifacts and, of course, subsistence strategies (cf. Kitcher, 1981, 1989; Strevens, 2004; Wylie, 2002).

I do admit that I may have somewhat exaggerated about how Korean archaeologists understand the relationship between Chulmun and Mumun. Beside, we have to consider the possibility of sudden and rapid transition from the former to the latter which caused the actual disconnection between the two at least in certain areas by human migrations (J. S. Kim, 2003, 2006). Anyhow, we have observed the connection between Chulmun and Mumun through the examination of changes in pattern on the potteries excavated from the central part of the Korean Peninsula (cf. Figure 2.8). In fact, there are archaeologists who have already recognized the similarities between the patterns on the potteries from the two periods (Shin, 2007). Nevertheless, many archaeologists in Korea still excavate and investigate their sites focusing on the differences between the overall archaeological assemblages of the two periods, and mostly adopt the migration model (J. S. Kim, 2003, 2006) to justify the artificial units over the real variation.

Frankly speaking, creating units for the interpretation of archaeological phenomena is somewhat necessary. Just like the scales which show mass or length — kilograms, meters, inches — archaeological units are useful for analysts for documenting the variation across the real things (O'Brien & Lyman, 1998). However, we should be aware that archaeological records must be understood as a 'continuous' sedimentary process, and concentrations of artifacts are the products of numerous events of deposition. Only with this awareness, units (e.g. Chulmun and Mumun) can be used as means of measurement.

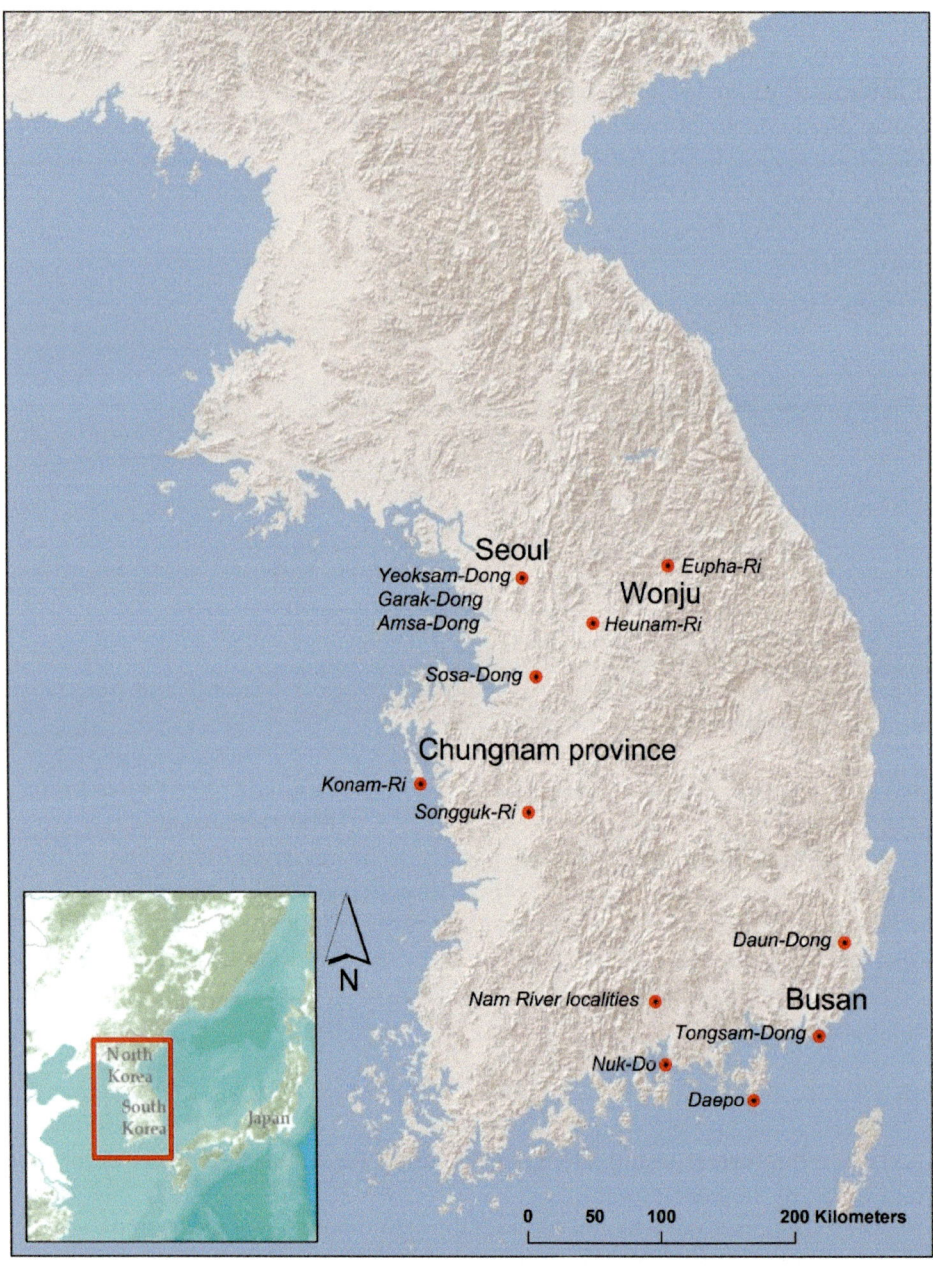

Figure 2.9: Location of the sites mentioned in the text

3. Methodological background, Research design and analytical procedure of the Luminescence dating

3.1. Luminescence dating in archaeology

To evaluate my hypothesis and to establish a general chronology of subsistence from 3,400 to 2,000 BP, I used organic geochemistry and luminescence dating methods on the pottery excavated from three major inland sites in the central part of the Korean Peninsula. In the Korean archaeology, the pottery is one of the main objects for the archaeological analysis, being abundant in the Korean Peninsula in almost every archaeological assemblage in sites that post-date 6,000 BP. This abundance has allowed archaeologists to develop a detailed Korean archaeological chronology based on the pottery shape, size and decoration. Though this intensive chronology-building has much contributed to the Korean archaeology, almost no attention has been given to analyzing the fabric of pottery itself. This is a surprising omission and represents a serious gap in our understanding of the prehistoric technology and subsistence. The above methods allow us to identify what was stored and cooked in the pots as well as to date them directly, so that we can understand how subsistence changed over time. Accordingly they let me directly test the hypothesis posited in the previous chapter: that there was utilization of a wider range of resources among ancient farmers in the central part of the Korean Peninsula between 3,400 and 2,000 BP and rice seems to have played no more than a minor role in subsistence during this period.

In terms of the pottery chronology, archaeologists have used stratigraphy that indicates depositional events: when the artifacts were buried together, not specifically when they were manufactured. Dating these depositional events or "occupations" (Dunnell, 1971; Rafferty, 2008) is a usual goal but it is not quite same as dating manufacturing events. Archaeologists have not always distinguished occupational events and manufacturing events in practice (cf. Feathers, 2009). In addition to stratigraphy, another method employed by archaeologists was the seriation based on the physical characteristics of the potteries. However, this also has an inherent problem, because transmission of the physical characteristics can occur across space (Dunnell, 1970; Feathers, 2009). To ascertain that the seriation is mainly entangled in time, it must be restricted to space. The lack of control over the spatial variation means it is difficult to tell whether there are sequential or special differences between each stage of a seriation. However, in real world archaeology, restricting the spatial variation is not always an easy task, especially when the research area is relatively large. The radiocarbon dating somewhat fitted with those traditional approaches, for this well-known absolute dating method mostly does not date the pottery themselves but nearby organic remains (e.g. Charcoal). This means the dating event inevitably has a variable relationship to the target event of pottery manufacture.

Luminescence dating dates the manufacturing event: when the pottery was made. To understand the chronology of subsistence, what archaeologists need to know is the age of the cooking event. Since the cooking event is more likely associated with the manufacturing event than with the depositional event, luminescence dating is probably the most suitable method for creating subsistence chronology. Luminescence dating provides a robust *terminus post quem* for the cooking event, in a way that is not possible using radiocarbon dating of organic remains.

3.2. Luminescence: The principals

Luminescence dating is an absolute dating method that has been used both intensively and extensively in the field of archaeology and Earth sciences. It is based on the emission of light, luminescence, from minerals. In case of pottery, burnt flints, or burnt stones, the dated event is the last heating of the objects. Another common application is dating sediments. In this case, the event being dated is the last exposure of the mineral grains to light. The age range to which the method can be applied is from a century or less to over one hundred thousand years.

Luminescence dating utilizes the radioactive isotopes of elements such as uranium (U), thorium (Th) and potassium (K) (Feathers, 2003). Radioactivity is ubiquitous in the natural environment. Naturally occurring common minerals such as quartz and feldspars act as dosimeters, showing the amount of radiation to which they have been exposed (Duller, 2008). A common characteristic of these naturally occurring minerals is that when they are exposed to the energy emitted by radioactive decay, they tend to store some proportion of it within their crystal structure. The minerals accumulate this energy as their exposure to radioactive decay continues through time. When this energy is released at some later date, it takes the form of light. This light is what we call luminescence.

Luminescence is explained by the solid state energy band theory (Aitken, 1985, 1998; McKeever & Chen, 1997). The interaction between radiation and the crystal structure provides energy to electrons that can be raised from the valence band to the conduction band. Because of this stage, electrons become trapped within the crystal. In the ideal situation, electrons cannot be trapped within the crystal structure, but their trapping is possible because of defects within the structure. Electrons may be stored (and accumulated) at these defects for a certain period. By the time these electrons are released, they lose the energy delivered by the radiation, and may emit a part of that energy in the form of a single photon of light (Duller, 2008).

The reason why we can use this phenomenon for dating lies in the fact that this energy stored in minerals can be reset by two processes. The first process is heating the material to the temperature above about 500°C: the process that occurs in a hearth or kiln during firing of pottery. The second is exposure to daylight, as may occur during erosion, transportation, or deposition of sediments. Either of these processes releases any existing energy, and thus set the 'clock' to zero (Duller, 2008). Therefore, in the luminescence dating, the event being dated is the last resetting of this clock, either by heat or light.

Measurement of the brightness of the luminescence signal can be used to calculate the total amount of radiation that the sample absorbed during the period of burial. If this is divided by the amount of radiation that the sample receives from its surroundings per year, it will give the duration of time for which the sample has been receiving energy: the age (Duller, 2008).

$$\text{age} = \frac{\text{total amount of radiation exposed during burial (equivalent dose)}}{\text{Amount of radiation receive each year (dose rate)}}$$

There are a number of naturally occurring minerals that emit luminescence signals, including quartz, feldspars, and calcite. Among them, quartz and feldspar are the most suitable and ubiquitous material for dating (cf. Feathers, 2003, 2009). The luminescence age is the period of time that has passed since the sample was heated or exposed to daylight. The age is given as the number of years before the date of measurement. Since there is no designate datum for luminescence ages, the date of measurement must be noted. The term BP (before present) should never be used for luminescence ages, for BP designates the specific datum point and is only proper for radiocarbon ages. The energy that is stored within minerals' crystal structure can be released using a number of laboratory methods.

3.2.1. Thermoluminescence

Heating the sample at a certain rate from the room temperature up to 700 °C releases the trapped electrons within the crystal structure. The resulting signal from this process is called thermoluminescence (hereafter TL). Typically the TL signal comes with a series of peaks (Figure 3.1). Each peak may indicate a single type of trap within the mineral, and commonly the signal comprises several traps. Although it is not always possible to identify the source of electrons precisely, in most cases TL signal observed at the highest temperature originates from the trap that is deepest below the conduction band (more energy is required to release electrons from deeper traps, and therefore this occurs at higher temperature).

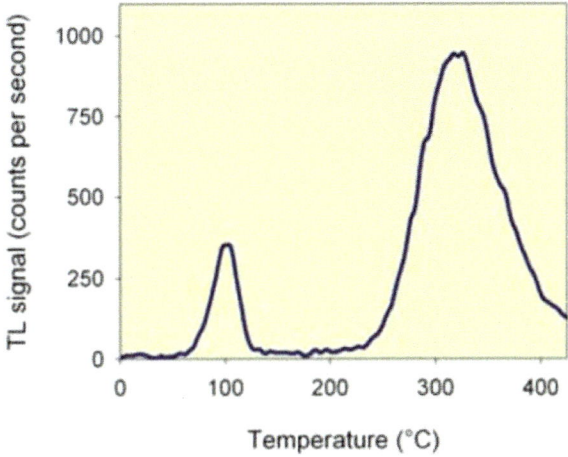

Figure 3.1. A typical thermoluminescence signal (commonly referred to as a "glow curve") that shows multiple traps (Duller, 2008; cf. Feathers, 2003: 1495)

3.2.2. Optically stimulated luminescence

A second way of releasing the electrons stored within minerals is exposing them to the laboratory light (Huntley et al.,1985). As soon as the mineral is exposed to light, the luminescence is emitted from the its grains. The signal is termed optically stimulated luminescence (hereafter OSL) and Figure 3.2 shows the signal from quartz during the stimulation. As the measurement continues, the electrons in the traps are emptied away and the signal starts to decrease drastically (Figure 3.2).

A similar signal is observed from other minerals including feldspar. However, OSL signal from feldspars decreases more slowly than that from quartz (Duller, 2008). Unlike TL, OSL signal does not shows multiple traps. Thus, before measuring the luminescence signal, it is important to thermally pretreat the sample to make sure that the measured signal comes from the deepest traps. This is achieved by heating the sample before measurement so that the shallow traps (whose electrons are unstable over the burial period) are emptied, leaving only the electrons in deeper, stable traps - this heating is called a preheat (Duller, 2008: 6; Feathers, 2003).

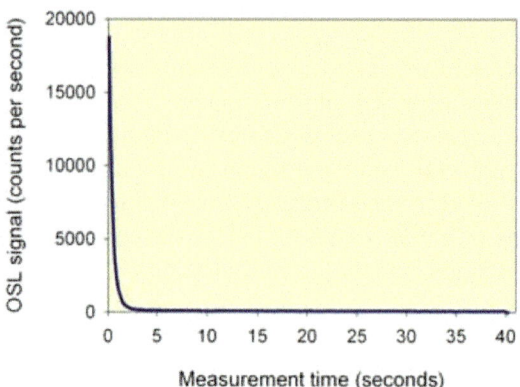

Figure 3.2. A typical optically stimulated luminescence signal from quartz grains (Duller, 2008)

3.3. Limits of the luminescence dating

3.3.1. Resetting of the signal

The archaeological value of the age obtained from the luminescence dating is determined by whether the resetting event is related to the archaeological event of interest. This means the investigator has to carefully consider the possibility of insufficient exposure or other exposures during the post-depositional process. For heated materials, the most crucial issue to consider is whether the sample was heated to a temperature high enough, and for a period of time long enough, for the trapped electron population to be completely removed. For unheated samples (mostly sediments), the important factors are the time and intensity of the light to which they were exposed during the depositional process. Inadequate exposure to daylight leaves a residual population of trapped electrons.

3.3.2. Accuracy and precision

Limitations on the precision of luminescence ages have been mentioned. When uncertainties in the measurement of the dose rate (often the issues related to the water content) and equivalent dose are combined, errors on luminescence ages normally range from 5 to 10%, including both random and systematic sources of error (Duller, 2008). In archaeological settings, understanding the archaeological context of the site and linking the date with it is a truly important factor for increasing both precision and accuracy. Barnett (Barnett, 2000) used the TL dates of the pottery from later prehistoric Britain to define the typological framework for that period. She found that where a diagnostic form and surface decorations were present, the correlation between the luminescence ages of potteries and the ages from other independent methods was high.

3.3.3. Upper and Lower age limits

The upper and lower age limits to which the luminescence dating is applicable vary from one place to another, and normally depend on the characteristic of the luminescence signal and does rate. The upper age limit is generally governed by the saturation of luminescence signal. At first, the luminescence signal increases almost linearly, but at some point the traps within the crystal structure where electrons can be stored become full. From here, the luminescence signal grows more slowly, until all the traps become full. When the luminescence signal ceases to grow despite continuous exposure, this is what we call saturation. This saturation determines an upper limit of the luminescence dating. Since the point at which saturation can be observed varies from one sample to another, it is impossible to give the precise upper limit to the age that can be obtained. Related to this, Wintle and Murray (2006) suggest that it is reasonable to work in the range where the natural signal is 85% or less of the maximum luminescence signal obtainable. Just as the upper limit, the lower age limit is also difficult to define. The lower limit is mostly controlled by two factors: (1) how well the luminescence signal was reset at the time of the event being dated and (2) the luminescence sensitivity of the mineral being studied.

3.4. Luminescence dating and its application to the Korea archaeology

The luminescence dating is a technique for dating once-heated or -exposed to sunlight materials, and is used by archaeologists primarily to date ancient ceramics and sediments (Feathers, 2003). This technique can measure the time that has elapsed since the last exposure to heat and light of the materials constituting the object. As this exposure event generally occurred when the pottery were made, the luminescence dating is ideal for dating archaeological ceramics (Feathers, 2003). The optically stimulated luminescence dating (hereafter OSL), infrared stimulated luminescence dating (hereafter IRSL), and thermoluminescence dating (hereafter TL) methods employed for dating ceramics have been quite common in Europe and the United States for nearly two decades, but they are yet to be widely used in Korea. Given the abundance of ceramics in Korean archaeological records, it is surprising that the luminescence technique has not been more frequently employed. Though it has been mentioned considerably since its initial introduction (J. H. Choi et al., 2006; J. C. Kim et al., 2009), it has been used mainly in the field of geology (Bang et al., 2009). In archaeology, after its applicability was considered (D. G. Hong et al., 2001), it has been employed to date several archaeological features including Bronze Age sediments (H. S. Lim et al., 2007), Paleolithic sediments (J. C. Kim et al., 2010), historic hydroponic farm (D. G. Hong et al., 2003), and potteries from the historic Three kingdom period (D. G. Hong et al., 2001; M. J. Kim et al., 2012). Probably the scarcity of archaeological luminescence dating in Korea may be attributed to the uncritical acceptance of the relative chronologies. I partially agree to the detailed relative chronologies based on the decoration and style of potteries and their serviceable nature (J. S. Bae, 2007; H. W. Lee, 2008). However, since these typological datings tend to ignore spatial variation, their accuracy could therefore be compromised in any particular location. In this regard, the typological dating has its uses, but the verification using luminescence is a prudent approach.

Lab. No	Depth (m)	Water Content (%)	Dose rate* (Gy/ka)	De (Gy)			Age
				TL	OSL	IRSL	
U2516	0.36	20.4	5.532±0.277	8.712±0.91 / 11.665±1.423	8.586±0.331	7.215±0.361	280±86 AD

Table 3.1: The result of the luminescence dating on the proto-historic period potsherd (*The dose rates are rounded to two decimal places, but the calculation of the total dose rate was carried out prior to rounding)

Of course, the primary purpose of the luminescence dating in this research is to investigate the role of the intensive rice farming and to establish the chronology of subsistence strategies over time by correlating the dates it obtained with the results of the organic geochemical analysis. However, with a systematic application of the luminescence dating, I was also able to grasp a glimpse of a more reliable chronology which can be easily applied to other archaeological studies. In 2011, I dated one potsherd from the archaeological deposit in Hongseong city, central part of the Korean Peninsula. Using the thermoluminescence method, I was able to confirm that the potsherd was from the proto-historic period (280 ± 86AD; U2516 in Table 3.1).

All the samples for my research was dated at the Luminescence Dating Lab, Department of Anthropology, University of Washington, under the direction of Dr James A. Feathers. The luminescence dating method enables the evaluation of the time that has passed since the mineral grains were last exposed to daylight or heated to a few hundred degrees Celsius. Generally, as at the lab of the University of Washington, the method uses an optically and thermally sensitive light or luminescence signal emitted by minerals such as quartz and feldspar. For dating, the amount of absorbed energy (luminescence signal) per mass of mineral (1 J/kg= 1 Gray) due to the natural radiation exposure since the last zeroing - known as the equivalent dose - is determined by comparing the natural luminescence signal of the sample with that which is induced by the artificial irradiation (Preusser et al., 2008). The time having passed since the last daylight exposure/heating (the date of the sample) is obtained through dividing the palaeodose by the dose rate, the latter representing the amount of energy deposited per mass of mineral by the radiation exposure on the sample over a certain time (Preusser et al., 2008). The potsherds in this study were dated by using this formula, and all the three methods, TL, OSL, and IRSL were applied. For a further clarification, the dates from the luminescence dating were correlated with those from AMS radiocarbon dating.

3.5. Analytical procedure

The luminescence dating method was developed in an archaeological context, in Europe in the 1960s and 1970s, as a method of dating heated materials, primarily ancient ceramics and potteries (Feathers, 2003). It has been applied to a wide range of Quaternary researches such as those on landscape evolution, palaeoclimate, archaeology, and has been being refined since its early days. It dates the past exposure to heat and light, and because the events of this exposure are the actual events archaeologists are interested in, it has a strong merit over other dating methods (Feathers, 2003). In other words, in the luminescence method, the dating event is often the target event that archaeologists are looking for. In this study, the luminescence dating was applied to seven archaeological ceramic samples.

3.5.1. Sample preparation - grain size

For the luminescence dating, determining the grain size is quite important, for it occasions diverse advantages/disadvantages as well as different methods. Generally, fine grains (1-8 um) are more abundant than coarse ones; and they can be analyzed with samples of relatively small amount. They also require a relatively simple sample preparation process, and rely less on the external dose rate, which is often problematic in a complex ceramic environment. However, if samples include feldspar grains (which cannot be separated from other grains during the sample preparation procedure), one has to deal with the high fading rate of feldspar (Wintle, 1973).

One of the biggest advantages of using coarse grains (180-212 um) is the single grain analysis, which can be done only with coarse grains. Quartz grains are generally used for the analysis of coarse grains, because of their well-known properties and low fading rate. Since it is possible to minimize feldspar inclusion during the sample preparation process of coarse grains, we do not have to consider the fading of feldspar as a major variable. Also, because of the larger grain size and etching process during the sample preparation, the contribution of alpha radiation (which has a short range: 50um) is minimal. This is a huge merit, for alpha radiation is much less effective in producing luminescence than beta and gamma radiations. In case of analyzing fine grains, this 'low alpha efficiency' must be considered. However, using coarse grains for the analysis requires a complicated sample preparation process and a larger amount of samples. Also, it cannot be totally exempted from the high fading rate, because feldspar has to be used for the single grain analysis in some cases (feldspar typically has a bright luminescence signal, which enables dating older deposits than with quartz) where quartz shows an extremely low luminescence signal (Preusser et al., 2008). It has also been verified that the quartz of volcanic origin may show anomalous fading, just like feldspar (Bonde et al., 2001; Tsukamoto et al., 2007). In this study, fine grains were used for the analyses, because of their small sample size and advantages that I have mentioned above.

3.5.2. Glassware and reagents

All glassware was washed with Decon 90 (Decon laboratories), rinsed four times in distilled water. Analytical grade reagents (typically ≥ 98% purity) were used throughout.

3.5.3. Dose rate measurement

The dose rate is the amount of energy deposited per mass of the mineral by the radiation exposure of the sample over a certain time (Preusser et al., 2008). For the dose rate measurement, the exposed parts of the potsherds were used (0.5-1 g). The dose rates were determined by alpha counting (Low level alpha counter 7286: Little more Science Engineering Co., DayBreak alpha counter 583: DayBreak), beta counting (Beta multi counter system RISØ GM-25-5: Risø National Laboratory), and flame photometry (Flame Photometer PFP-7: Jenway).

The water absorption percentages of the samples were measured. This is quite important for calculating the dose rate, as the attenuation of radiation is much greater if the sample is filled with water (Preusser et al., 2008). For measuring the water absorption percentage, the sample was saturated with deionizing water for several days. Then, its surface wetness was removed by gently dabbing it with a wet paper towel; and then it was immediately placed on the scale to weigh it. After the sherd was dried in a 50 °C oven for several days to record its weight in its dry state. The water absorption percent is calculated as $W = [(S/D)/D]*100$, where S is the saturated weight and D, the dry weight.

Some component of the dose rate is produced by the ionizing cosmic radiation, and could be different by the geographic location and burial depth of the sampled material (Prescott & Hutton, 1994). All infor mation related to the latter points was obtained from the excavation records of the sites where the samples came from. Alpha counting gives the current alpha activity rate. And based on this rate and the assumption of secular equilibrium, one can calculate the beta and gamma dose rate. However, by using the beta counter and flame photometry as well, we can enhance the validity of the total dose rate measurement (flame photometry is used to measure K content and the beta counter is used to assess theaccuracy of alpha counting and flame photometry measurements). This sort of advantage is available only if we utilize multiple tools at the same time.

3.5.4. Equivalent dose measurements

For measuring the equivalent dose (paleodose) of the pottery samples, TL (Thermo luminescence; Day-Break 11000 Automated TL system), OSL (Optically stimulated luminescence; RISØ TL/OSL system DA-15), and IRSL (Infrared stimulated luminescence; RISØ TL/OSL system DA-15) were utilized. Artificial laboratory irradiations were given by the Irradiator type 721/A (Little more Science Engineering Co.) and RISØ TL/OSL system DA-15. For beta radiation, Sr-90/Y-90 beta source, calibrated against a Cs-137 gamma source, was used. Am-241 source was used for Alpha irradiation. Fine grains (1-8 um fractions) were used for dating. The grains were obtained from the core part of the potsherds more than 2 mm away from any exposed surface. This was done by drilling, using tungsten carbide drill bits.

For the TL analysis, the equivalent dose was determined by the slide method to obtain both of the advantages of the additive dose method and the regeneration method (Aitken, 1985; Prescott et al., 1993). The slide method can deal with the matter of extrapolation as well as the change in sensitivity simultaneously. These two problems cannot be solved at the same time in case of using either the additive dose method, or the regeneration method solely. The regeneration curve can be used to define the extrapolated area and can be corrected for sensitivity change by comparing it with the additive dose curve. The equivalent dose is taken as the horizontal distance between the two curves after a scale adjustment for sensitivity change.

OSL and IRSL on fine-grain (1-8μm) pottery samples are carried out on a single aliquot following procedures adapted from Banerjee et al. (2001) and Roberts & Wintle (2001). The equivalent dose is determined by the single-aliquot regenerative dose (SAR) method (Murray & Wintle, 2000). The SAR method measures the natural signal and the signal from a series of regeneration doses on a single aliquot. The method uses a small test dose to monitor and correct for sensitivity changes brought about by preheating, irradiation or light stimulation. SAR consists of the following steps: (1) preheat, (2) measurement of the natural signal (OSL or IRSL), (3) test dose, (4) cut heat, (5) measurement of test dose signal, (6) regeneration dose, (7) preheat, (8) measurement of the signal from regeneration, (9) test dose, (10) cut heat, (11) measurement of the test dose signal, (12) repeat of the steps from 6 to 11 for various regeneration doses. Usually a zero regeneration dose and a repeated regeneration dose are employed to insure the procedure is working properly. For fine-grained ceramics, a preheat of 240 °C for 10s, a test dose of 3.1 Gy, and a cut heat of 200 °C are currently being used, although these parameters may be modified from sample to sample.

For OSL and IRSL, the luminescence was measured on a RisØ TL-DA-15 automated reader by a succession of two stimulations: first 100 s at 60 °C of IRSL (880nm diodes), and then 100s at 125 °C of OSL (470nm diodes). Detection is effected through 7.5mm of Hoya U340 (ultra-violet) filters. The two stimulations are used to construct IRSL and OSL growth curves, so that two estimations of equivalent dose are available. Feldspar usually involves anomalous fading and only feldspar is sensitive to IRSL stimulation. The rationale for the IRSL stimulation is to remove most of the feldspar signal, so that the subsequent OSL (post IR blue) signal is free from anomalous fading (Roberts & Wintle, 2001). However, feldspar is also sensitive to blue light (470nm), and it is possible that IRSL does not remove all the feldspar signal. Some preliminary tests in our laboratory suggested that the OSL signal does not suffer from fading, but this may be sample specific. The procedure is still undergoing study.

As I mentioned above, for dating fine-grained samples, one has to deal with the low alpha efficiency. This is taken into account by determining the alpha efficiency factor: "b-value" (Huntley et al. 1988). It has been known that the alpha efficiency varies between quartz and feldspar (Huntley et al. 1988). The typical b-value of quartz and feldspar is respectively about 0.5 and more than 1.5. For TL, the alpha efficiency is determined by comparing additive dose curves using alpha and beta irradiations. The slide program is also used in this regard, taking the scale factor (which is the ratio of the two slopes) as b-value (Aitken, 1985). The results from several samples from different geographic locations show that OSL b-value is less variable and centers around 0.5. IRSL b-value is more variable and is higher than that for OSL. TL bvalue tends to fall between the OSL and IRSL values. Currently, measuring the b-value for IRSL and OSL is in process by giving an alpha dose to aliquots whose luminescence have been drained by exposure to light. An equivalent dose is determined by SAR using beta irradiation, and the beta/alpha equivalent dose ratio is taken as b-value. A high OSL b-value is indicative that feldspar might be contributing to the signal and thus subject to anomalous fading.

3.5.5. Determining the age

The time having passed since the last daylight exposure/heating of the pottery sample (Hereafter: age) was calculated through dividing the palaeodose by the dose rate. The final date of the sample was obtained through calculating the average of the three dates from TL, OSL, and IRSL. Normally, when conducting the luminescence dating on a pottery sample, its associated sediment is required for the precise dose rate measurement. However, since there was no associated sediments on my samples, I relied on an average of sediment dose rates determined in other parts of Korea (D. G Hong et al., 2003; J. C. Kim et al., 2010; M. J. Kim et al., 2012; H. S. Lim et al., 2007). The age and error for both OSL and TL are calculated by a laboratory constructed spreadsheet, based on Aitken (1985). All error terms are reported at 1-sigma.

4. Methods, Research design and analytical procedure of the organic geochemical analysis

4.1. Concept of biomolecular archaeology and organic geochemical analysis

Biomolecular archaeology is the study of ancient biomolecules that can provide information relating to human activities in the past (Evershed, 2008b; Stear, 2008: 24). According to Stear (2008), the area of biomolecular archaeological researches includes (1) the use of collagen from skeletal remains to determine the ancient dietary information (Corr et al., 2008; J. J. Lee, 2011b; Reynard & Hedges, 2008; Richards et al., 2003; Thompson et al., 2008); (2) the analysis of DNA from archaeological materials to explore evolutionary origins and migratory patterns (Edwards et al., 2004; Ho et al., 2008; Jansen et al., 2002; Malhi et al., 2007; Vilà et al., 2001); and (3) the study of lipid biomarkers from a range of archaeological contexts relying on the organic geochemical analysis for the reconstruction of culinary, economic and social practices throughout prehistory and history (Berstan et al., 2004; Bethell et al., 1994; Buonasera, et al., 2015; Copley et al., 2005a, 2001; Craig et al., 2013, 2011; Dudd et al., 1998; Evershed et al., 1997; 2003; Hansel et al., 2004; Reber & Evershed, 2004b; Regert et al., 2003). Organic geochemical analysis endeavors to determine the types of food groups that were cooked or stored within a pot by attempting to isolate and identify the specific organic compounds trapped in the fabric of its wall or adhering to its surface in residues (Eerkens, 2002, 2005, 2007; Evershed et al., 1990; Reber & Evershed, 2004a). Organic compounds have the advantage that they are often preserved within archaeological ceramics (Charters et al., 1993; Copley et al., 2005a, 2005b; Evershed, et al., 1994; Heron & Evershed, 1993), which is not the case in the other methods of diet reconstruction, such as examination of faunal and floral remains. In this regard, the organic geochemical analysis has become an important method of investigation which archaeologists use to better understand local diets and the function of ceramic artifacts. If we conduct it on pottery, we will be able to understand past subsistence behaviors in relation to pots even in the absence of faunal or floral remains. The direct examination of remains of organic resources in the Korean Peninsula has typically been limited to shell middens, because the high acidity of sediments does not allow long-term preservation of bone or plant remains. Therefore, organic geochemical analysis is a suitable method to investigate organic resources in non-midden sites in Korea.

4.2. Organic Residues within archaeological potteries

Among all the compound classes I have mentioned above, solvent-extractable lipids are the most frequently recovered compounds from archaeological contexts (Evershed, 1993, 2008a, 2008b). Because of their stability against degradation and inherent hydrophobicity, they tend to persist at the original place of deposition more than other biomolecules. Due to these characteristics, lipids are nowadays the most widely studied organic compounds in the discipline of biomolecular archaeology.

Under favorable conditions, lipids are preserved at archaeological sites in association with a wide range of archaeological contexts, e. g. potteries, sediments, human and animal remains (Evershed, 1993; Evershed et al., 1999; Mukherjee, 2004). Among them, potsherds are probably the most widely distributed at archaeological sites. Due to this reason, the pottery is one of the most extensively studied material cultures for the organic geochemical analysis.

Organic residues are found in association with archaeological potteries either as (1) charred remains on the inner or outer surface of vessels, or, (2) absorbed within the fabric of their wall (Evershed, 2008b; Evershed et al., 1999). The residues both on their surface and in their fabric can provide invaluable information regarding the use of ancient pottery vessels. However the latter case is more commonly encountered, for the fired clay acts as a 'trap' or 'net', protecting and preserving lipids during burial

(Evershed et al., 2001; Reber & Evershed, 2004a). Studies have shown that these compounds are relatively well insulated and preserved within that fabric over millennia (Eerkens, 2001, 2005; Heron et al., 1991). The absorbed residues, unlike the visible ones, cannot be removed from a sample by washing or scraping, and remain within the ceramic matrix of the pot until extracted by solvents (Reber & Evershed, 2004a: 20).

During the usage of pottery vessels in prehistoric times (e.g. during culinary practices), fats, oils and waxes originated from animals, insects or plant products become entrapped within the vessel wall. The fats and waxes are protected from microbial and chemical degradations as well as groundwater leaching by the ceramic matrix. These organic residues can be extracted from potsherds and analyzed hundreds or even thousands of years after the pottery was discarded by ancient people. For example, in case of Great Britain, absorbed residues are typically detected in 50 to 60 % of all the vessels studied (Mukherjee, 2004); however, the actual proportion is dependent on many factors including burial conditions and age (Evershed et al., 2008). Though Fats and waxes can also be preserved in the form of charred or dried deposits adhering to the vessel wall, this class of residue is much less commonly observed.

The preservation of organic compounds in the porous wall of the pottery was first recognized over 30 years ago, when the lipids extracted from archaeological potteries were analyzed by the gas chromatography (hereafter GC) (Condamin et al., 1976; Stear, 2008: 25). This approach uses the ratio between the amounts of common fatty acids to determine particular classes of food (cf. Eerkens, 2005, 2007; Patrick et al., 1985). But it has a problem, for different kinds of fatty acids decompose at different rates over time due to oxidation and hydrolysis. Since such ratios are not stable over time, researchers have to rely on those of the fatty acids that decompose at similar rates. For example, Eerkens (2001, 2005, 2007) set up the criteria for distinguishing different food classes, based on four useful ratios involving eight fatty acids which are relatively common in archaeological residues ($C12:0/C14:0$, $C16:0/C18:0$, $C16:1/C18:1$ and $(C15:0 + C17:0)/C18:0$). Upon these criteria, he was able to distinguish five different food classes which are: meat of terrestrial mammals, fish, seeds/nuts and berries, roots, and greens (Table 4.1). After these studies that attempted to determine the origins of the organic residues based on the proportions between individual compounds, more sophisticated mass-spectrometric instruments were employed and made it possible to identify a wide range of organic commodities within archaeological vessels.

The identification and characterization of lipid residues rely upon the comparison of chemical properties of lipid compounds derived from organisms. Those compounds are presented in both the archaeological ceramics and contemporary plants and animals. Such "biomarkers" can help scientists to reconstruct the dietary life of prehistoric peoples (Evershed, 2008a; Evershed et al., 2008; Heron & Evershed, 1993: 267-270). This is achieved by the high temperature gas chromatography (hereafter HTGC) and gas chromatography - mass spectrometry (hereafter GC-MS) techniques that can acquire detailed molecular compositional information from the extracts. That information can subsequently be compared to that of modern reference materials. Through this method, scholars have identified terrestrial and marine animal fats, plant leaf waxes (e.g. cabbage and leek), beeswax, birch bark tar, and palm fruit (Table 4.2). But the biomarkers only occur in case of good preservation of the organic residues; more often we only have the degraded products. More recently, the use of soft ionization techniques in MS, such as electrospray ionization (ESI), has proven particularly useful in the structural characterization of high molecular weight compounds preserved within the archaeological pottery like triacylglycerols (hereafter TAGs). They are more difficult to examine with the GC-MS technique (Mirabaud et al., 2007; Stear, 2008: 26).

ratio	State	terrestrial mammals	fish	Roots	greens	seeds/nuts and berries
C16:0/C18:0	Fresh	<3.5	4–6	3–12	5–12	0–9
	degraded	<7	8–12	6–24	10–24	0–18
C12:0/C14:0	Fresh	<0.15	<0.15	>0.15	>0.05	>0.15
	degraded	<0.15	<0.15	>0.15	>0.05	>0.15

Table 4.1: Criteria used to distinguish food types, based on fatty acid ratios (Eerkens 2005)

Most recently, the application of the compound-specific stable carbon isotope analysis (hereafter CSIA) by the gas chromatography-combustion-isotope ratio mass spectrometry (hereafter GC-C-IRMS) enabled a more specific characterization of the organic compounds within the archaeological pottery. The stable carbon isotope analysis has become a powerful method for tracing diet patterns of animals, for the isotopic composition of animals depends upon the food they eat (Malainey, 2011). In archaeological settings, the method has been widely used on human remains for understanding human subsistence patterns by distinguishing C_3 diets (e.g. rice) from C_4 diets (e.g. millet) (Barton et al., 2009; Bentley et al., 2007). In the field of ceramic studies, Hastorf and DeNiro (1985) conducted the bulk carbon isotope analysis for charred organic residues on the surface of potsherds to understand human diets. With the introduction of

Commodities	Lipid biomarkers	References
Terrestrial animal fats	Characteristic distribution of TAGs, diacylglycerols (hereafter DAGs), monoacylglycerols (hereafter MAGs) and free fatty acids. Particularly high abundance of C16:0 and C18:0 fatty acids.	Evershed et al., 2001
Marine animal fats	Isoprenoid fatty acids (4, 8, 12-trimethyltridecanoic acid and phytanic acid). Thermally produced ω-(o-alkylphenyl)alkanoic acids	Hansel et al., 2004; Coplay et al., 2004; Craig et al., 2011
Plant waxes (e.g. brassica wax)	Long chain alcohols, ketones, n-alkanes, aldehydes and wax esters. Specific biomarkers of brassica wax (cabbage) nonacosane, nonacosan-15-ol, nonacosan-15-one.	Evershed et al., 1991
Beeswax	Characteristic distribution of odd number n-alkanes (C23-C33), even numbered free fatty acids (C22-C30), and long chain palmitic wax esters (C40-C52)	Evershed et al., 1997; Regert et al., 2003a
Birch bark tar	Triterpenoids from lupane family, namely betulin, lupeol and lupenone	Charters et al., 1993
Palm fruit	High abundance of C12:0 and C14:0 saturated fatty acid	Coplay et al., 2001a, b

Table 4.2: Identification of fatty acids by using GC-MS (Stear 2008: 26)

GC-C-IRMS, the stable carbon isotope value of individual compounds in a mixture can now be measured with high precision, providing a unique opportunity to conduct the carbon isotopic analysis on the fatty acids that are insulated within the fabric of archaeological ceramics (Mottram et al., 1999). Scholars have been successfully tracing the presence of C_3, C_4 plants, animal fats, and aquatic resources (e.g. fish and mammals) on prehistoric potsherds through CSIA (Craig et al., 2013, 2011; Cramp et al., 2011; Evershed et al., 1994, 1997; Mottram et al., 1999; Reber & Evershed, 2004a; Salque et al., 2013).

4.3. Identification of lipids

Different criteria can be used for the identification of lipid residues. For example, the presence of fatty acids can indicate a plant or animal origin through their relative abundance, while the TAG distribution and structure are also potentially useful indicators (Mukherjee, 2004). However, caution must be exercised when using these criteria, for ratios between fatty acids may change over time and TAGs are often only present in very low abundance or completely absent (Mukherjee, 2004: p. 14). In addition, because of the differential degradation and variable extraction rate of organic compounds, it is hard to tell exactly what types of food were processed in the pot only with the GC-MS analysis (cf. Reber & Evershed, 2004b). A more reliable method for the elucidation of the lipid origin is to determine the stable carbon isotope (hereafter $\delta^{13}C$) value of individual C16:0 and C18:0 fatty acids.

In this study, I have conducted the organic geochemical analysis on the absorbed lipids extracted from the potsherds. The analysis involves two different analytic methods: GC-MS and CSIA based on GC-C-IRMS. The former is used for separation and identification of organic compounds within a potsherd, and the latter can be employed for the further isotopic analysis of specific compounds. If fatty acids such as C16:0 and C18:0 are found in a range of different food products, the isotopic analysis can further distinguish between their origins. Most of the recent organic geochemical studies on potsherds successfully detected the presence of different food groups including animal fat, ruminant milk, marine resources (e.g. fish and mammals), fresh water resources, C_3, and C_4 plants with those two methods combined (Craig et al., 2011; Cramp et al., 2011; Reber & Evershed, 2004a).

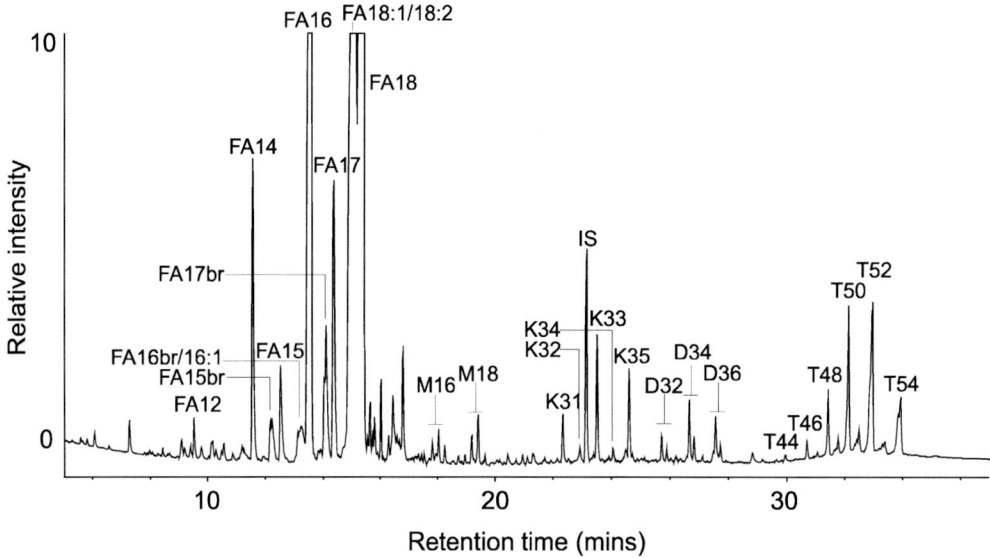

Figure 4.1: partial HTGC profile of the lipid extract from a Romano-British sherd from Stanwick, Northamptonshire (Evershed et al., 2002: 661). A low abundance of intact TAGs are observed at retention times above 30 min. The majority of them was hydrolyzed during vessel use or burial, resulting in the formation of DAGs, MAGs, and free fatty acids. IS (internal standard: n-tetratriacontane) was added to the sample at the extraction stage for quantification of lipid. The extracts are trimethylsilylated.

4.3.1. GC-MS analysis

GC-MS enables the identification of even highly degraded commodities. A reliable classification of commodities processed in the archaeological pottery can be made by comparing the chemical structure of individual compounds with that of modern and archaeological references (Mukherjee, 2004). A knowledge of the degradative process occurring during vessel use and burial is essential in order to identify the lipid residues preserved within vessels. These analyses are enhanced by analyzing the results of laboratory and field experiments simulating use and degradation (Dudd & Evershed, 1998; cf. Dudd et al., 1998; Evershed, 2008a).

Figure 4.1 shows an example of degraded animal fat obtained by HTGC analysis of a Romano-British sherd from Stanwick, Northamptonshire (Mukherjee, 2004: 14). A low abundance of intact TAGs are observed at retention times above 30 min; however, the majority of the lipid was hydrolyzed during vessel use or burial, resulting in the formation of DAGs, MAGs, and free fatty acids. The fatty acids present, eluted between 10 and 20 min, comprise mainly C16:0 and C18:0 components. A high abundance of C18:0 is indicative of animal fat.

Distributions of TAGs in ancient fats from pots can provide a reasonable evidence for the presence of animal fats and dairy products (Mukherjee, 2004: 19). For the detection of TAG 'biomarkers', GC-MS is used, which can help to make distinction between different kinds of animal fats (Dudd & Evershed, 1998). For example, bovine adipose fats possess saturated TAGs of every carbon number between C44 and C54 and pig fats contain a narrow distribution of them (e.g. TAGs range from C46 to C54) (cf. Mukherjee, 2004). On the other hand, milk fats are quite distinctive because of their relatively wide TAG distribution ranging from C40 to C54 (Dudd et al., 1998; Evershed et al., 2003). Figure 4.2 shows TAG distributions of both fresh/degraded lipid residues gathered from the modern reference fats. Most importantly, however, it should be addressed that distributions of TAGs alone are not sufficient enough for the proper identification of lipid origin (Mukherjee, 2004). Moreover, TAGs frequently do not survive in archaeological residues. Due to this vulnerable characteristic, sometimes TAGs may be misinterpreted. Figure 4.2d and e indicate fresh and degraded ruminant milk fat. Since the degradation process during vessel use or burial makes the ruminant milk TAG distribution (4.2e) similar to those of adipose fats (4.2a; b), a more prudent decision has to be made based on a more robust stable isotopic criterion (Copley et al., 2003; Dudd et al., 1998; Mukherjee, 2004: 20).

In this study, GC-MS was applied to identify the compounds which are only found in certain food groups (cf. Table 4.2). The biomarkers which these compounds constitute are present in different types of fats; for example, short chain fatty acids in dairy fat, unsaturated fatty acids in plant oil, cholesterol in animal fats and plant sterols (e.g. b-sitosterol) in plant oil. Especially, Phytanic acid (3,7,11,15-tetramethylhexadecanoic acid) and 4,8,12-TMTD (4,8,12-trimethyltridecanoic acid) are isoprenoid compounds which are mostly found in particularly high concentrations in marine animals (Evershed, 2008b). Along with thermally produced long-chain ω-(o-alkylphenyl)alkanoic acids, these compounds are indicators of aquatic/marine resources (Craig et al., 2011; Evershed, 2008b). But, as I already indicated, they only occur in case of good preservation of food residues. One way to deal with this preservation issue is to use GC-MS in the selection monitoring (SIM) mode, where the analysis focuses on specific biomarkers, in order to try to get a better signal from the compounds which may be present in very low quantities, or which may be masked by more abundant compounds such as C16:0 and C18:0 fatty acids.

4.3.2. Compound specific isotope analysis

In most cases a pot is reused over time, and may be used to cook different kinds of food from one cooking episode to another. Researches with amino acids show that the first use of a pot essentially saturates it with them, and seals it off further amino acid contributions, that is, the amino acid residues trapped within a pot record only its first use (Fankhauser, 1997). On the other hand, fatty acids and other compounds tend to accumulate in the fabric of the pot wall. Therefore, the result of the analysis is, in this case, more likely to reflect the entire usages of the pot. Generally, the result is assumed to represent the type of food group that was most frequently processed in it. However, this does not mean we can just disregard the complication caused by its multiple usages. Besides, due to the differential degradation and variable extraction rate of the organic compounds, it is not easy to tell exactly what types of food were processed in the pot only with GC-MS analysis (cf. Reber & Evershed, 2004b). On top of that, animal fats and plant oils offer a great challenges, because their major components, unsaturated fatty acids in particular, rarely if ever, survive, leaving mainly rather undiagnostic n-alkanoic acids such as C16:0 and C18:0 fatty acids (derived mainly through the hydrolysis of triacylglycerols, Figure 4.3, Evershed et al., 2008).

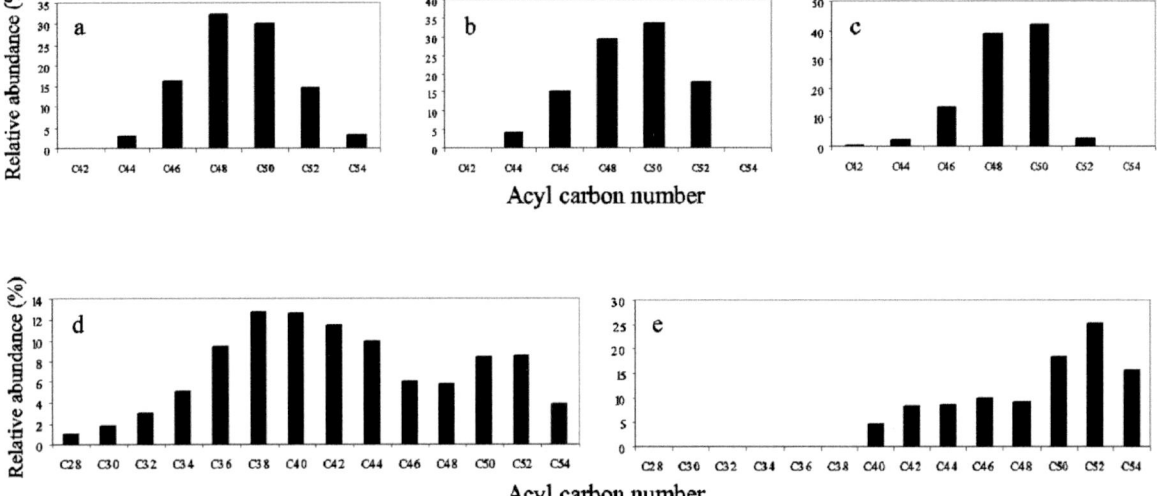

Figure 4.2: The distributions of TAGs in different kinds of animal fats (modified from Mukherjee, 2004: 20). (a): cow adipose fat (b): sheep adipose fat (c): pig adipose fat (d): fresh milk (e): milk degraded for 90 days

Luckily, we do have the last approach that can help us further clarify the origin of the organic compounds in a pot: compound specific stable carbon isotope analysis. Early works of stable isotope study in the archaeological field involved the bulk isotopic analysis (Hastorf & DeNiro, 1985; Morton & Schwarcz, 1988). However, the application of CSIA via GC-C-IRMS allows us to achieve a greater specificity, for the structure of diagnostic compounds in complex mixtures can be directly linked to their stable isotope value (Evershed et al., 1994). Thus, the compound specific stable isotope analysis avoids ambiguities arising from contamination by, e.g. plasticizers originating from plastic bags in which sherds are often stored. These ambiguities cannot be resolved in the bulk isotope analysis (Mukherjee, 2004). Most importantly, you do not need to have solid materials (e.g. bone) for the analysis.

Generally, different food groups tend to have different major fatty acids having different ranges of $\delta^{13}C$ values (e.g. C16:0 and C18:0). For example, $\delta^{13}C$ values of ruminant (goat, sheep and cow/buffalo), chicken, equine, pig fat, ruminant milk, C_3 plant, C_4 plant, and aquatic resources (e.g. fish and mammals), have each their own range. Therefore, $\delta^{13}C$ values of fatty acids provide the basis for distinguishing those food classes. Though these values were obtained from the modern fauna and flora, they have been employed as references for many archaeological studies (Craig et al., 2011; Cramp et al., 2011; Fraser et al., 2012; Reber & Evershed, 2004a, 2004b). In proceeding in this fashion, these studies assume that the δ13C values of modern samples are comparable to those of ancient members of the same species. Scholars were able to detect the presence of the above classes of food by measuring $\delta^{13}C$ values of the two most common fatty acids in archaeological pots: palmitic acid (C16:0) and stearic acid (C18:0), with GC-C-IRMS, which provides a means to address some key questions concerning human subsistence in prehistory (Craig et al., 2013; Evershed et al., 1994, 1997; Mottram et al., 1999; Salque et al., 2013).

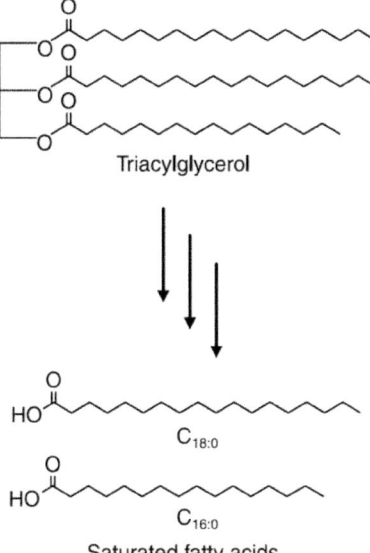
Figure 4.3: Undiagnostic C16:0 and C18:0 fatty acids generated through the hydrolysis of triacylglycerols due to the degradation of fat/oil during burial process. As biomarkers, C16:0 and C18:0 fatty acids have a severely limited diagnostic value (Evershed, 2008b: 900).

In nature, carbon exists as three isotopes: ^{12}C and ^{13}C, which are both stable, and 14C, which is radioactive. Occurring as CO_2 (carbon dioxide), they are organizing respectively 98.89 %, 1.11 %, and 1 x 10^{-10} % of the global carbon pool (Stear, 2008). Being inorganic, carbon dioxide is incorporated into living organisms through the process of photosynthesis. Green plants transform carbon dioxide and water into oxygen and organic sugars. When incorporated into the plant tissue through photosynthesis, the isotopic fractionation occurs and the ratio between ^{13}C to ^{12}C changes significantly, because plants use the carbon dioxide containing the lighter isotope, $^{12}CO_2$, more readily than that of the heavier isotope, $^{13}CO_2$. Plants are consumed by herbivores, and herbivores are consumed by carnivores. If one can measure the ratio between ^{13}C to ^{12}C in the remains of those organisms and compare it with known reference isotope ratios, then it will be possible to trace their diet.

The stable carbon isotope ratio is measured by comparing the relative differences of ^{13}C to ^{12}C between the sample and the international standard, Pee Dee belemnite (PDB), a limestone from South Carolina (Malainey, 2011):

It is expressed using the delta (δ) notation:

$$\delta^{13}C = \left(\frac{R\ sample - R\ Standard}{R\ Standard}\right) \times 1000$$

Where:
R sample = molar $^{13}C/^{12}C$ ratio of the sample,
R standard = molar $^{13}C/^{12}C$ ratio of the standard

The $\delta^{13}C$ value is the difference between the 13C content of the sample and that of the standard, and is expressed relatively to the international standard. Differences between samples are very small, so values are counted per mil (‰), rather than percent (%). The standard contains less ^{12}C and more ^{13}C than most natural materials, so $\delta^{13}C$ values of the samples are usually negative, ranging between -37 and -8 ‰. The error range for compound specific $\delta^{13}C$ values of fatty acids is ± 0.3 ‰.

4.3.2.1. Modern reference animal fats and plant oils

Naturally, plants and animals of today cannot be directly compared to those of prehistoric times, due to the various environmental changes that have occurred over the last few hundred years (Mukherjee, 2004). There are several factors of these changes including: (1) consuming fossil fuel since the industrial revolution which has caused changes in the isotopic composition of CO_2 in the air (Friedli et al., 1986); (2) commercial farming due to which animals have been fed with supplements to enhance their diets and to improve the nutritional quality of their meat and milk (cf. Chilliard et al., 2001; Lowe et al., 2002); and (3) selective breeding that has introduced changes in the composition of the fat and milk of domestic animals (Mukherjee, 2004: 17). There are also regional level factors. For example in Great Britain, since C_4 plants (e.g. millet) have been introduced and incorporated into animals' diet not long ago, it is hard to directly compare $\delta^{13}C$ values of modern and prehistoric animals (Stear, 2008).

The identification of plant oils through the isotope analysis is possible, for the range of $\delta_{13}C$ values is different in each group of plants that share the photosynthetic pathway. Terrestrial plants use three different photosynthetic pathways, namely C_3, C_4 and CAM. The C_3 plants (e.g. wheat, rye, barley, legumes) are the most abundant, and are found mainly in moderate areas. They fix the atmospheric CO_2 using the Calvin and Benson cycle (Calvin & Benson, 1948). $^{13}CO_2$ is discriminated by Ribulose-1,5-bisphosphate carboxylase/oxygenase (hereafter RuBisCO), resulting in relatively low $\delta^{13}C$ values ranging from -32 to -20 ‰ (Boutton, 1991). C_4 plants (e.g. millet, maize, sugarcane, sorghum) fix CO_2 through the Hatch-Slack pathway (Hatch & Slack, 1966), and the carbon fixation occurs near the surface of the leaf in mesophyll cells with phosphoenolpyruvate (hereafter PEP). The latter pathway gives relatively high $\delta^{13}C$ values in the range of -17 to -12.5 ‰ (Malainey, 2011). Crassulacean acid metabolism (hereafter CAM) plants (e.g. pineapple, aloe vera, jade plant) can either assimilate CO_2 at night only or night and day. The carbon fixation occurs at night through PEP carboxylase as in C_4 plants. On the other hand, during the day time, CAM plants can switch their photosynthetic pathway and use RuBisCO to fix CO_2. As a result, the range of ^{13}C values for some CAM plants is quite broad (cf. Malainey, 2011).

For the identification of animal fats originated from the archaeological pottery, they were compared with the carefully assembled data of modern fats (Copley et al., 2003; Craig et al., 2013; Dudd et al., 1998; Evershed et al., 2003). The treatment of modern fats to create the reference database is slightly different from case to case. In Britain, only the animals that are being reared on known diets were sampled in order to form the database (e.g. C_3 plant diet in order to mimic the prehistoric condition, absence of C_4 plant), which includes adipose fats from cattle, sheep and pigs, and milk fat from cattle and sheep (Copley et al., 2003; Dudd et al., 1998; Evershed et al., 2003). The $\delta^{13}C$ values from these animals reflect their different diets and variations in their metabolism as well as physiology (Evershed et al., 1999; Stear, 2008). The ellipses shown in Figure 4.4a indicate the $\delta^{13}C$ values obtained from the C16:0 and C18:0 fatty acids from each of the reference animal fats; sheep and cattle data are grouped together as ruminant fats. Dairy and adipose fats from ruminant animals can be distinguished, for the C18:0 fatty acid in dairy fat is significantly more depleted in $\delta^{13}C$ value (average 2.1 ‰, Copley et al., 2003). In Japan, to avoid the effects of commercial farming and selective breeding, modern reference samples were collected from authentic wild animals (Figure 4.4b). To facilitate comparison with archaeological data, the $\delta^{13}C$ values obtained from all modern reference animals were adjusted by the addition of 1.2 ‰ considering post-Industrial Revolution effects of fossil fuel burning (Friedli et al., 1986).

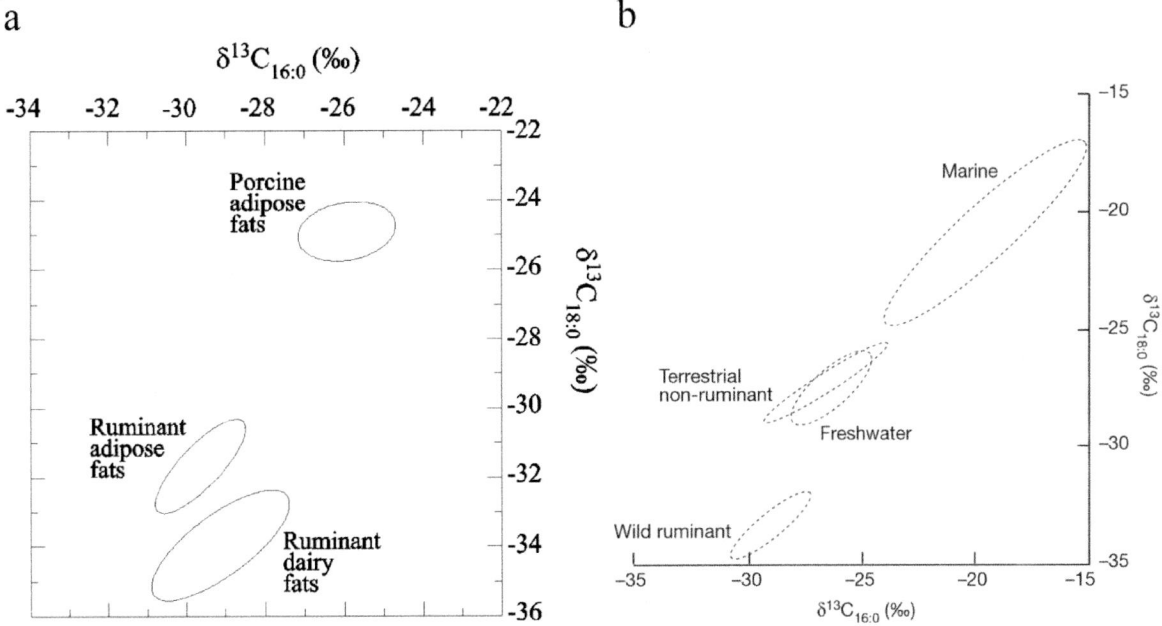

Figure 4.4: Reference database created based on modern fats for CSIA. (a): Only the animals having been reared on known diets were sampled (e.g.C₃ plant diet in order to mimic the prehistoric condition, absence of C₄ plants) (Copley et al., 2003; Dudd et al., 1998). (b): The modern reference samples were collected from authentic wild animals to avoid the effects of commercial farming and selective breeding (Craig et al., 2013). The δ¹³C values obtained from all modern reference animals were adjusted by the addition of 1.2 ‰, considering post-Industrial Revolution effects of fossil fuel burning (Friedli et al., 1986).

4.3.2.2. Interpretation of CSIA

For the interpretation of CSIA, the $\delta^{13}C$ values acquired from the C16:0 and C18:0 fatty acids in archaeological potsherds are plotted in the figure of the reference animal fat ellipses (Figure 4.5a). When the $\delta^{13}C$ values of fatty acids plotted within an ellipse, like the case of the pork (porcine) fat in Figure 4.5a, then the fat in question can be identified as pork fat. When the ^{13}C values are plotted just outside the ellipse, then the fat can be identified 'predominantly' as pork fat (Mukherjee, 2004). However, in most cases the $\delta^{13}C$ values are located between the ellipses of the reference fats, which indicates the mixing of different classes of food stuffs within the vessel either at a moment or during all the time of its use (Mukherjee, 2004).

To account for the mixing of different animal fats in varying proportions within a single vessel, a theoretical mixing model is used to calculate theoretical $\delta^{13}C$ values (Bull et al., 1999; Mukherjee, 2004: 22):

$$\delta^{13}C_{mix} = \delta^{13}C_{(A)}\left(\frac{(X \times A)}{(X \times A)+(Y \times B)}\right) + \delta^{13}C_{(B)}\left(\frac{(Y \times B)}{(X \times A)+(Y \times B)}\right)$$

Where:
$\delta^{13}C_{mix}$ = predicted $\delta^{13}C$ value of the fatty acid with contributions from fats A and B
$\delta^{13}C_{(A)}$ = $\delta^{13}C$ value of the individual fatty acid in fat A
$\delta^{13}C_{(B)}$ = $\delta^{13}C$ value of the individual fatty acid in fat B
X = percentage of fat A present (%)
Y = percentage of fat B present (%)
A = percentage of the individual fatty acid in fat A (%)
B = percentage of the individual fatty acid in fat B (%)

Theoretical mixing curves between the porcine adipose fat, ruminant adipose fat and ruminant dairy fat are shown in Figure 4.5b. The ellipses which represent different food classes (ruminant adipose fat, ruminant dairy fat and porcine adipose fat) are connected by a theoretical mixing curve (Figure 4.5b).

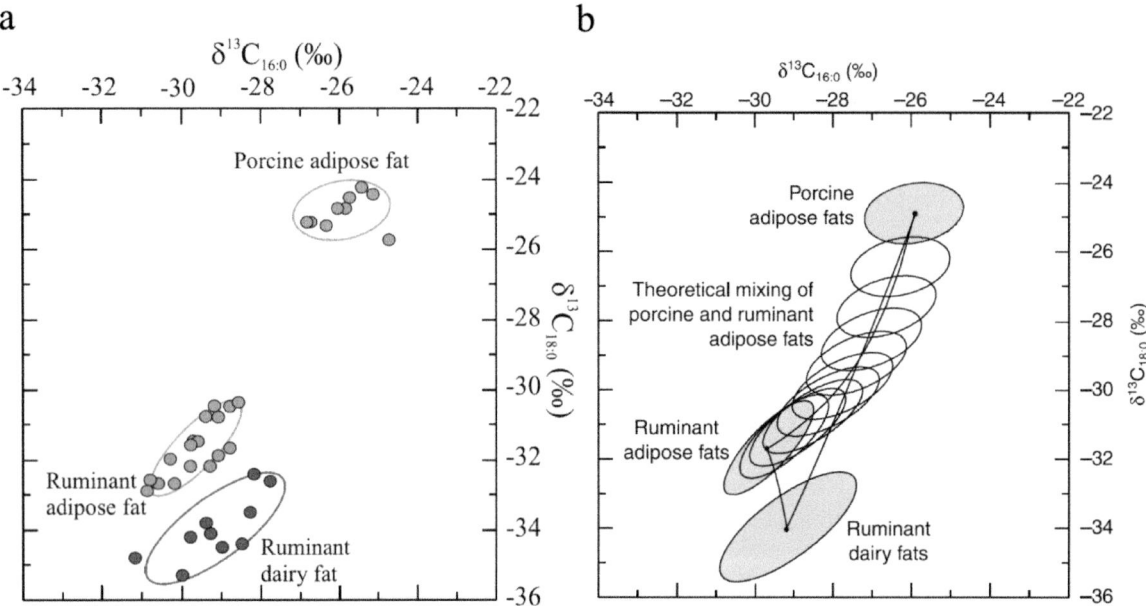

Figure 4.5: Interpretation of the results of CSIA (a): $\delta^{13}C$ values acquired from the C16:0 and C18:0 fatty acids in archaeological potsherds are plotted along with the reference animal fat ellipses (Evershed, 2007: 151) (b): The theoretical mixing curves between the porcine adipose fat, ruminant adipose fat and ruminant dairy fat are shown (Evershed, 2008b: 900)

When utilizing this theoretical mixing model for the interpretation of the contributions of different foodstuffs within a mixture, we need to consider several important points. First of all, it is nearly impossible to quantify exactly how much mixing was occurred during each vessel use, and how often each vessel was subsequently re-used (Mukherjee, 2004). It is also difficult to estimate the exact relative amount of different food classes cooked in a vessel over its lifetime usage, for the concentration of the fatty acids from different food classes varies significantly (Enser, 1991).

4.3.2.3. Possibilities of variation in $\delta^{13}C$ values of the fatty acids from the archaeological lipid

According to the Mukherjee (2004: 23-30), there are several possible sources that can affect $\delta^{13}C$ values of fatty acids from archaeological lipid.

4.3.2.3.1. Forest density and depletion of ^{13}C

In the areas covered with dense forest we see a significant deviation of 13C distribution from the global average causing plants to be depleted of ^{13}C. In these regions, a positive correlation between the forest density and the degree of depletion of ^{13}C is observed. In addition, there is a gradual variation of $\delta^{13}C$ values of tree leaves from the ground to the top of the tree; and it indicates that the most negative values occur near the ground (Medina & Minchin, 1980; Vogel, 1978). This is what we call the 'canopy effect' (Medina & Minchin, 1980). The average bulk $\delta^{13}C$ value of C_3 plants in open air areas is about -26 ‰. However, for the leaves in a subtropical monsoon forest, a $\delta^{13}C$ value of -35 ‰ was recorded, and a value as low as -37 ‰ was observed in the Amazon forest (Ehleringer et al., 1987; Medina et al., 1980).

This phenomenon in dense forest areas will influence the $\delta^{13}C$ values of fatty acids extracted from the local ruminant animals and pigs dwelling in forest (Van Der Merwe & Medina, 1989). Therefore, if the reference animals used for the study were not raised within the forest environment, they may have more enriched $\delta^{13}C$ values of fatty acid, compared with their ancient counterparts which dwelled in forest. That is, fatty acids from archaeological fats might indicate more negative $\delta^{13}C$ values than those of their modern counterparts; and this must be carefully considered.

4.3.2.3.2. Variations in $\delta^{13}C$ values of CO_2

Things change over time. Any variation in the atmospheric CO_2 which occurred over time as a result of a climate change or environmental fluctuation, may have caused a deviation of $\delta 13C$ values of archaeological animal fats from the reference values. The variation in $\delta^{13}C$ value of the atmospheric CO_2 from the multiplied tree-ring record obtained from oaks suggests it can vary up to 1.5 ‰ (Figure 4.7a), McCormac et al., 1994). Even within a relatively short term, the $\delta 13C$ value of the atmospheric CO_2 can vary quite dynamically (Figure 4.7b, Robertson et al., 1997). It is likely that other terrestrial plants will also show variations in a similar way, but their scale might differ between the species (Mukherjee, 2004). The differences in $\delta^{13}C$ values between modern and ancient fats resulting from such a temporal variation of the atmospheric CO_2 can be overcome by comparing $\delta^{13}C$ values of modern reference and archaeological fats.

4.3.2.3.3. Variation related to human activities

As mentioned above, the theoretical mixing curve was calculated to consider mixing of different food products within a single vessel during its lifetime usage. However, in some cases, mixture with other uneatable natural products is often observed. For example, the beeswax contained in lipid extracts from potsherds appears often as a mixture with degraded fats from foodstuffs (Mukherjee, 2004). Beeswax is characterized by a distribution of linear hydrocarbons of odd-numbered carbon (C21 - C23), free fatty acids of even-numbered carbon (C22 - C30), and/or long-chain wax esters with the carbon number range from C40 to C52 (Kolattukudy, 1976; cf. Mukherjee, 2004). Though the exact reasons for the presence of beeswax in archaeological pottery vessels are yet unknown, it may have been used as a 'slip' due to its hydrophobic characteristic, or it might be a byproduct of the use of honey in cooking/flavoring. The abundant C16:0 fatty acid present in modern beeswax exhibits a $\delta^{13}C$ value of around -26.4 ‰, while the C18:0 fatty acid is present in low abundance (Mukherjee, 2004). In this situation it is important to assess whether there is a significant isotopic contribution from natural products like beeswax and how it may influence our interpretation of isotopic analyses.

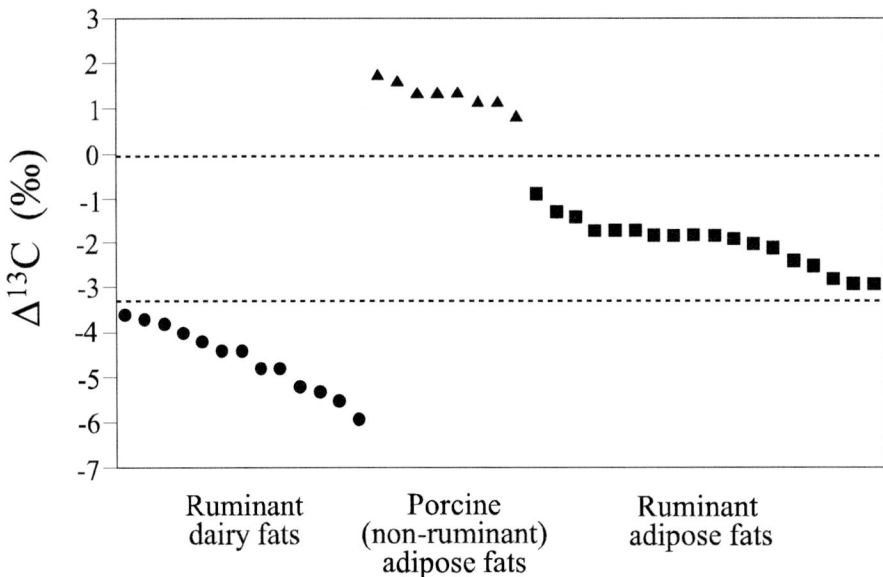

Figure 4.6: Plots showing the difference in $\delta^{13}C$ values of the C18:0 and C16:0 fatty acids ($\Delta^{13}C = \delta^{13}C_{18:0} - \delta^{13}C_{16:0}$) obtained from the modern reference fats (Copley et al., 2003: 1526)

When conducting CSIA, the best way to establish the reference database is to collect modern samples of fauna and flora from the same region where archaeological materials were collected. However, as I mentioned above, the modern day's commercial farming with supplements makes it impossible for us to directly compare the $\delta^{13}C$ values from archaeological materials with those from modern samples. To overcome this issue, scholars have been collecting samples from wild fauna and flora for creating the reference database. Unfortunately, in case of Korea, since wild terrestrial mammals are extremely rare, it is beyond the scope of this study.

In this study, as for the CSIA, the archaeological samples from the central part of the Korean Peninsula were sent to the Stable Isotope facility at the University of California-Davis, and analyzed by Varian CP3800 GC coupled onto a Saturn 2200 ion trap MS/MS. Based on the results, the stable carbon isotope values of C16:0 and C18:0 fatty acids from the archaeological samples will be compared with the available modern references that were obtained from the modern fauna and flora that exist in either Japan, Northern Europe or North America (Copley et al., 2003; Craig et al., 2013, 2011; Cramp et al., 2011; Dudd, et al., 1999; Dudd et al., 1998; Evershed et al., 1994, 1997; Mottram et al., 1999; Reber & Evershed, 2004a; Steele et al., 2010) to detect the presence of the potentially cooked resources in the prehistoric Korean Peninsula. Since the overall ecosystem of Japan, Northern Europe, and North America is similar to that of Korea and almost all the fauna and flora having produced the data for reference exist also in the Korean Peninsula, this approach assumes that the $\delta^{13}C$ values of available modern samples are comparable to archaeological ones from the Korean Peninsula.

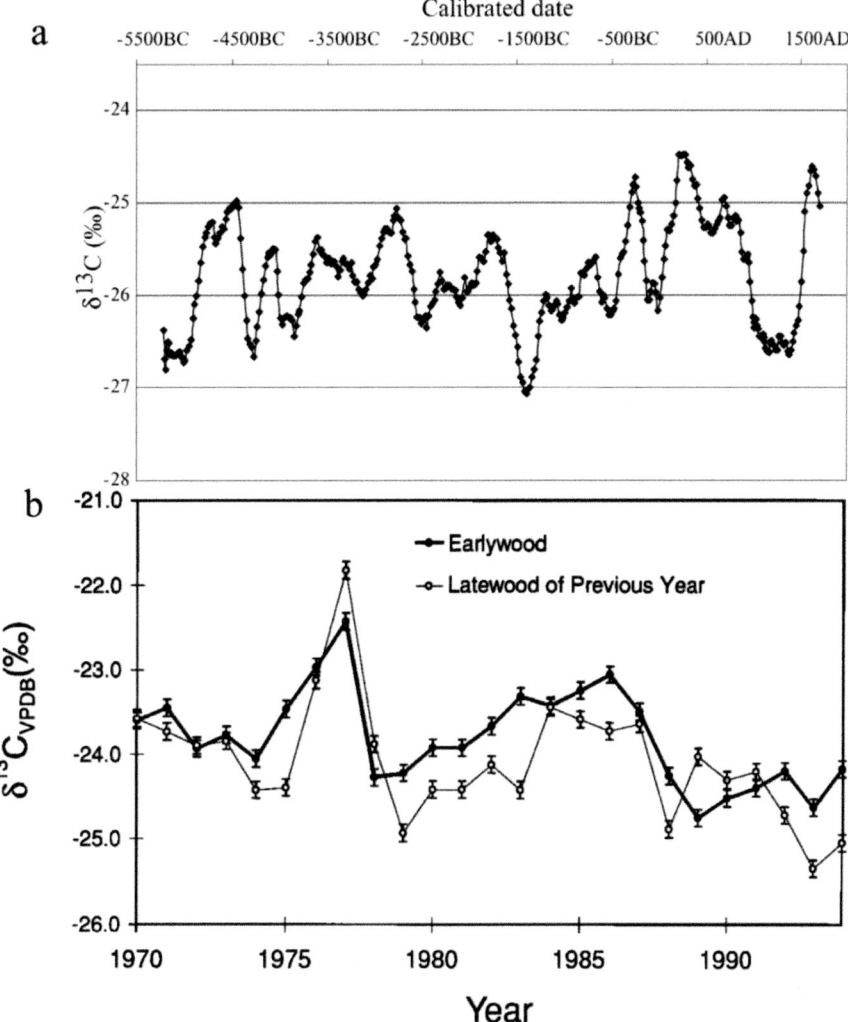

Figure 4.7: $\delta^{13}C$ values of the cellulose from the oak tree-ring sequence (a): 11-year running mean from ancient Irish oaks (data obtained from McCormac et al., 1994; Mukherjee, 2004: 28) (b): Yearly measurement from 1970 to 1995 of modern oaks in east England (Robertson et al., 1997)

4.4. Analytical procedures

Lipids are medium-sized molecules which possess predominantly linear, branched or cyclic hydrocarbon skeletons making them soluble in organic solvents (Correa-Ascencio & Evershed, 2014). For this reason, the most well-known way of the extraction of organic compounds is using a solvent mixture (e.g. chloroform-methanol 2 : 1 v/v) and the ultra-sonication of powdered potsherds. The main purpose of this approach is to extract free fatty acids and other organic compounds which are absorbed and trapped in the voids of clay matrixes. This way of extraction of lipids from archaeological ceramics by a solvent mixture has proven its effectiveness in different parts of the world. However, Craig and his colleagues (2004) showed that the lipid recovery can be incomplete when extracting with a solvent mixture, and some portions of residues do remain non-extractable without the use of a stronger extractant (e.g. methanolic sodium hydroxide). As a response to that, Correa-Ascencio & Evershed (2014) recently developed a new extraction protocol which uses acidified methanol (2% sulfuric acid-methanol v/v). According to Correa-Ascencio & Evershed (2014), this new "methanolic acid extraction" has several advantages over the method of conventional solvent extraction:

(1) The new method can recover both free and bound lipids from the ceramic matrix and therefore, is especially effective in increasing the recovery rate of lipid residues from the archaeological pottery containing those of low concentration (Figure 4.9). In this regard, the application of this new method has the potential to expand the limits of the analysis of archaeological lipid residues when lipid preservation is limited.

(2) The simultaneous extraction and derivatization of lipid residues, for the further isotopic analyses, shorten significantly the examination time to one day of overall laboratory time instead of four to five days required when the chloroform : methanol extraction method is applied; and they also shorten the require of materials.

(3) The major disadvantage of the new method is the compositional information loss due to the hydrolysis of complex lipids (e.g. acylglycerols and wax esters) during the extraction process. However, the loss of these lipids is not problematic, as they are the components that occur rarely, or in very low abundance, in most archaeological assemblages.

In this study, both methods were employed to test their suitability for the Korean Peninsula. Figure 4.8 shows the differences between the solvent and acid extractions.

4.4.1. Glassware, solvents and reagents

All the solvents used for this research were HPLC (High-performance liquid chromatography) grade. The reusable glassware were washed with Decon 90 (Decon laboratories), rinsed with acetone, dried in the oven at first and heated in the furnace (450 °C; 24 hours). In order to prevent contamination, combusted foil and tweezers were used to manipulate the samples. Analytical blanks were prepared with each batch of samples during each procedure of lipid extraction and derivatization to monitor any possible source of contamination. Analytical grade reagents (typically $\geq 98\%$ purity) were used throughout.

4.4.2. Solvent extraction of lipids

The lipids were extracted following an established protocol outlined in Figure 4.8a. Approximately 5-10g of each potsherd was sampled and its surface was cleaned using a drill (Dremel 3000) to remove any external contaminants, such as those originating from soil or fingers due to handling during the excavation/curation process. The cleaned sample was ground to fine powder in a glass mortar & pestle and accurately weighed to be put in a glass vial. The lipids were extracted using chloroform : methanol (2:1; 10 mL) and sonicated (20 min. × 2). The extract was then centrifuged (2500 rpm; 10 minutes.) and only the liquid portion containing the Total Lipid Extraction (hereafter TLE) was removed and transferred to a glass vial. The TLE was filtered through a silica column (1 g) to remove any particulate matter and accidental inclusions of solid materials. About a half portion of the TLE was derivatized to form Trimethylsilyl (hereafter TMS) ethers prior to analysis by GC-MS. The other half was derivatized to fatty acid methyl esters (hereafter FAMEs) and analyzed by GC and GC-C-IRMS.

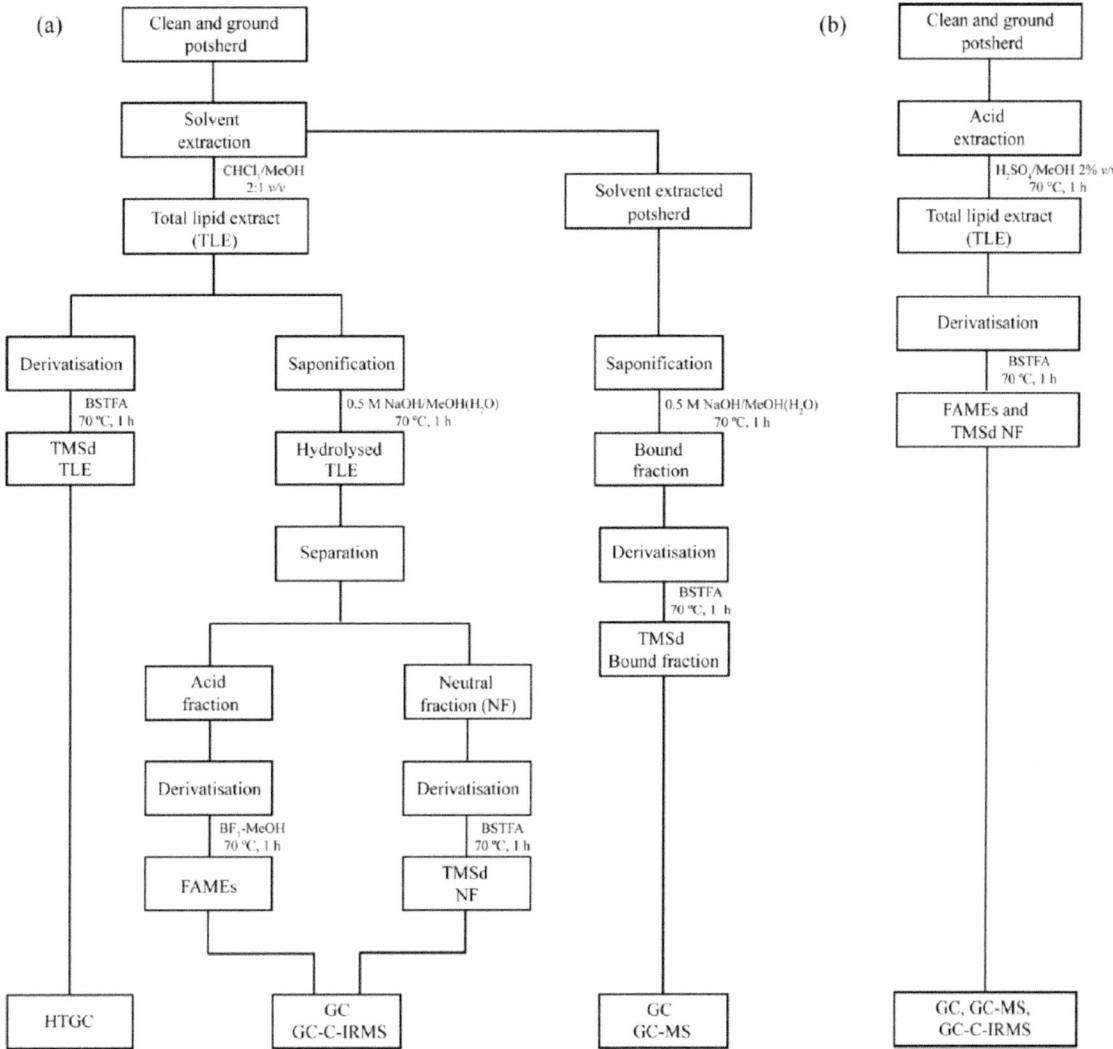

Figure 4.8: The comparison between (a) the solvent extraction protocol and (b) the acid extraction protocol (Correa-Ascencio & Evershed, 2014: 1331)

4.4.2.1. Preparation of TMS derivatives

One half of the TLE was treated with N,o-bis(trimethylsilyl)trifluoroacetamide (hereafter BSTFA) containing 1 % trimethylchlorosilane (40 uL; 70 °C; 1 hour). Then, BSTFA was removed under gentle nitrogen gas and the derivatized TLE was dissolved in toluene (50 uL) prior to GC-MS.

4.4.2.2. Preparation of FAMEs

The FAME derivatives of the free fatty acids were prepared by heating them with BF3-methanol (14 % w/v; 100 uL; 70 °C; 1 hour). Nano-purified water was added (1 mL) and the FAME derivatives were extracted with chloroform (3 × 2 mL) and the solvent was removed under nitrogen. The FAMEs were redissolved in hexane prior to the analysis by GC-MS and GC-C-IRMS.

4.4.3. Methanolic acid extraction of lipids

The lipids were extracted following an established protocol outlined in Figure 4.8b. Approximately 5g of each potsherd was sampled and its surface was cleaned using a drill (Dremel 3000) to remove any external contaminants. The cleaned sample was ground to fine powder in a glass mortar & pestle and accurately weighed. The sample was transferred into a culture tube (I) and 5mL of H_2SO_4 (sulfuric acid) : MeOH (methanol) were added to it; and the whole was heated (2% v/v, 70 °C, 1 hour, vortex-mixing every 5 minutes). It is important to check the pH after extraction to examine whether the sample is still acid, for carbonate- rich ceramic fabrics might neutralize acid. If the pH is ≥ 3, then more H_2SO_4 : MeOH should be added.

The H_2SO_4 : MeOH solution containing the extract was transferred to the test tube, and centrifuged for 10 minutes (2500 rpm). The clear solution was transferred to another clean culture tube (II) and 2mL of nano-purified water were added. Then, 4 mL of hexane were dropped in the culture tube (I), and vortexmixed to recover any lipids which are not fully extracted by the methanol solution. The hexane portion was transferred in the culture tube (II) and vortex-mixed with the H_2SO_4 : MeOH solution to extract the lipids. The washing of the culture tube (I) with hexane and vortex-mixing in the culture tube (II) were repeated twice. Then, the hexane portion was transferred to a clean vial. Following this, 2 mL of hexane were added directly to the H_2SO_4 : MeOH solution in the culture tube (II), and vortex-mixed with it to extract the remaining lipid residues. The hexane extracts were gathered in a clean vial, and evaporated under a gentle nitrogen blow, and re-dissolved in 300 uL of hexane for GC-MS and GC-C-IRMS.

4.4.4. Analysis with GC-MS and GC-C-IRMS

4.4.4.1. High Temperature GC-MS

The trimethylsilylated TLEs and FAMEs were analyzed by 6890N Network GC system with a 5979 Mass selective Detector from Agilent Technologies at the Sachs laboratory, Department of Oceanography, University of Washington. The GC was equipped with a fused silica capillary column (J&W; DB5-MS; 60m × 0.32 mm; 0.25 B5m film thickness) and the interface was maintained at 110 °C. The mass spectrometer was operated in the full scan mode. Helium was the carrier gas and the GC oven was programmed as follows: 2 min isothermal at 50 °C are followed by an increase to 350 °C at a rate of 10 °C min-1 and following this, the temperature is held at 350 °C for 10 min. The peaks are identified based on their mass spectral characteristics and GC retention times, and also by comparison with the NIST mass spectral library.

4.4.4.2. GC-C-IRMS

The CSIA Analysis was performed using a Thermo GC/C-IRMS system composed of a Trace GC Ultra gas chromatograph (Thermo Electron Corp., Milan, Italy) coupled onto a Delta V Advantage isotope ratio mass spectrometer through a GC/C-III interface (Thermo Electron Corp., Bremen, Germany). A compound identification support for the CSIA laboratory is provided by a Varian CP3800 gas chromatograph coupled onto a Saturn 2200 ion trap MS/MS (Varian, Inc., Walnut Creek, CA U.S.A.). The FAMEs dissolved in hexane were injected in the splitless mode, and separated on a Varian factor FOUR VF-5ms column (30m × 0.25mm ID, 0.25 micron film thickness). Once separated, the FAMEs are quantitatively converted to CO2 in an oxidation reactor at 950 °C. Following water removal through a nafion dryer, CO_2 enters the IRMS. The $\delta 13C$ values were corrected using the working standards composed of several FAMEs calibrated against the NIST standard reference materials. Each sample was analyzed ten times.

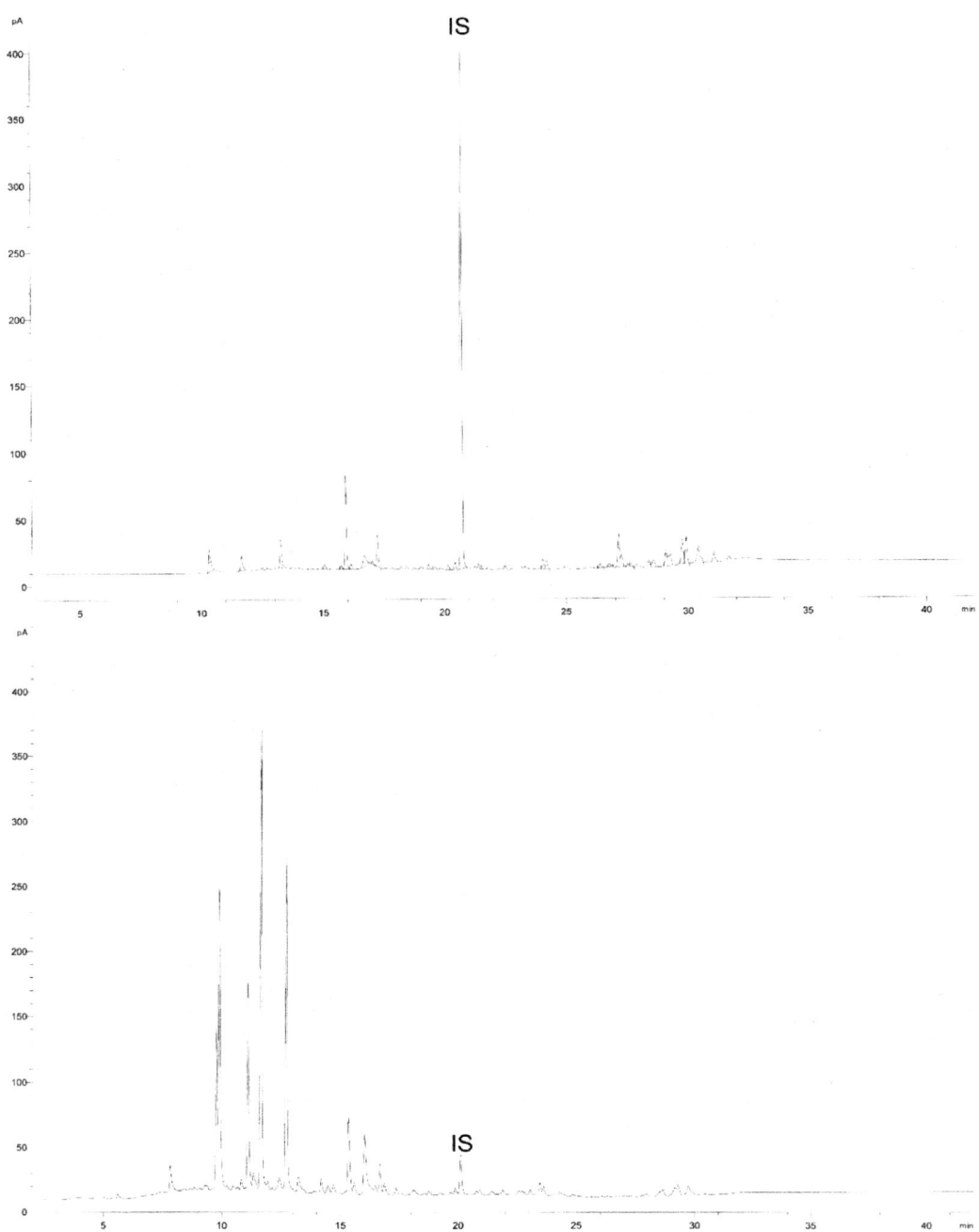

Figure 4.9: The GC chromatograms of the same archaeological sherd sample (KIM014) showing different recovery rates. (a) chloroform : methanol solvent extraction (b): acidified methanol extraction (IS = Internal Standard). In both extractions, the same amount of internal standard was injected. The acidified extraction method showed a much higher recovery rate (more than 20 times) compared with the prevailing chloroform/methanol solvent extraction protocol.

5. The Results

5.1. Kimpo-Yangchon

The Kimpo-Yangchon site is located on the low hillocks surrounded by Guree-Ri, Yoohyeon-Ri, and Yangchon-Ri of Kimpo city, Gyeonggi province. The site is about 4 kilometers southwest of the Han River (Figure 2.7; 5.2). The research period was from October 30th, 2007 to February 25th, 2011. The site includes various archaeological phenomena such as house pits, mound burials, pit graves, stone-lined pit burials, and firing features which represent different time periods from the Chulmun period to the historical Joseon Dynasty (AD 1392 - 1897) (B. M. Kim et al., 2013). The total area of the site is 863,992 square meters. Its main archaeological phenomena belong to the Mumun period, and the analysis was focused on this time period.

Six house pits and two pit features were classified into the Chulmun period. The house pits are either round shaped or square shaped with rounded corners, and hold the interior features such as hearth, post holes and ditch. Most of the potteries are pointed-bottomed deep bowls with various combinations of patterns including (short) slanted incising, herringbone, and lattice. The excavated house structures are assumed to belong to the late Middle - Late Chulmun period.

As for the Mumun period, 126 house pits, pit features, and firing features were excavated. The house pits are classified into three types based on their shape: square, rectangular, and longhouse. Each of those houses normally has an array of multiple post holes which crosses the center of the pit; and some of them contain pit-hearths, storage pits, and ditches as interior features. Most of the potteries have the rim-punctuation or a combination of lip-scoring/rim-punctuation; and others a combination of doublerim/short slanted line incision (Figure 2.4). As for the ground stone tools, arrowheads, daggers, and axes were found. As for the farming tools, semi-lunar shaped stone knives (Figure 5.4; cf. Figure 2.6b) and mortar/pestle were found. The excavated features can be reclassified into two different lineages: (large) square/rectangular house pits with double-rim/short slanted line incision potteries and (small) rectangular house pits/(elongated) long houses with rim-punctuation potteries (Figure 2.4). These two lineages are considered to be an extension of the two Early Mumun pottery cultures (Garak-Dong style and Yeoksam-Dong style) which covers a large extent of Gyeonggi province (Figure 2.7). These Mumun features of the site have a great value in understanding the overall aspect of the Mumun period in the central west part of the Korean Peninsula. Considering the number of houses, artifacts (Figure 5.4), and the radiocarbon dating on the charcoal from hearths in the house pits (B. M. Kim et al., 2013, Table 5.1), the period when the Kimpo-Yangchon site was occupied the most intensively is around 3,000 - 2,700 BP, the incipient/early stage of the Mumun period (cf. Figure 5.1; 6.3).

Location/house pit No.	Cultural historical period	C^{14} date (BP; uncalibrated)	Calendar date
Area 2-1 "B"/No.1	Mumun	2650±50	BC 815
Area 2-1 "B-1"/No.1	Mumun	3010±50	BC 1255
Area 2-1 "B-1"/No.2	Mumun	2540±40	BC 770
Area 1-D /No.22	Mumun	2770±60	BC 910
Area 1-D /No.23	Mumun	2850±40	BC 1005
Area 2-1 "F"/No.1	Chulmun	4530±50	BC 3175
Area 2-1 "F"/No.2	Chulmun	4550±50	BC 3175
Area 1-G /No.4	Mumun	2700±40	BC 835
Area 1-G /No.2	Mumun	2680±50	BC 830
Area 1-H /No.5	Mumun	2380±40	BC 455
Area 1-H /No.12	Mumun	2770±40	BC 935
Area 2-1 "B-1"/No.3	Mumun	2950±50	BC 1175
Area 2-1 "J"/No.1	Mumun	2670±40	BC 820
Area 2-1 "J"/No.3	Mumun	2820±50	BC 975
Area 2-1 "J"/No.4	Mumun	2740±50	BC 875
Area 2-1 "J"/No.6	Mumun	2830±50	BC 980
Area 2-1 "J"/No.9	Chulmun	4020±50	BC 2525
Area 2-1 "J"/No.10	Mumun	2710±40	BC 860
Area 2-1 "J"/No.12	Mumun	2900±50	BC 1100
Area 2-1 "J"/No.13	Mumun	2650±50	BC 815
Area 2-1 "J"/No.13	Mumun	2920±50	BC 1130
Area 2-1 "J"/No.16	Mumun	2630±40	BC 808
Area 2-1 "J"/No.18	Mumun	2560±50	BC 775
Area 2-1 "K"/No.1	Mumun	2900±50	BC 1100
Area 2-1 "K"/No.2	Mumun	2630±40	BC 808
Area 1-K /No.3	Mumun	3020±50	BC 1300
Area 1-L /No.3	Mumun	2960±50	BC 1190
Area 1-L /No.5	Mumun	2750±50	BC 885
Area 1-L /No.6	Mumun	2550±40	BC 770
Area 1-L /No.10	Mumun	2820±50	BC 975
Area 1-L /No.11	Mumun	2910±50	BC 1105
Area 1-L /No.12	Mumun	2820±60	BC 975
Area 1-L /No.13	Mumun	2750±50	BC 885
Area 1-L /No.14	Mumun	2800±50	BC 955
Area 1-L /No.15	Mumun	3090±60	BC 1360
Area 1-L /No.16	Mumun	2990±50	BC 1215
Area 1-L /No.17	Mumun	2910±50	BC 1105
Area 1-L /No.19	Mumun	2720±50	BC 863
Area 1-L /No.20	Mumun	2550±50	BC 770
Area 2-3 "Na" /No.1	Mumun	2520±50	BC 595
Area 2-3 "Na" /No.3	Mumun	2660±60	BC 845
Area 2-3 "Na" /No.4	Mumun	2760±50	BC 885
Area 2-3 "Na" /No.6	Mumun	2680±56	BC 830
Area 2-3 "Na" /No.7	Mumun	2710±50	BC 858
Area 2-3 "Na" /No.8	Mumun	2920±50	BC 1130
Area 2-3 "Na" /No.15	Mumun	2850±50	BC 1010
Area 2-4 "Ga" /No. 2	Baekje Kingdom	1730±80	AD 320
Area 2-4 "Ga" /No. 11	Baekje Kingdom	1670±50	AD 375
Area 2-4 "Ga" /No. 13	Baekje Kingdom	1670±60	AD 375
Area 2-4 "Ga" /No. 8	BaekJe Kingdom	1880±60	AD 145

Table 5.1: The results of the AMS radiocarbon dating of the Kimpo-Yangcho site

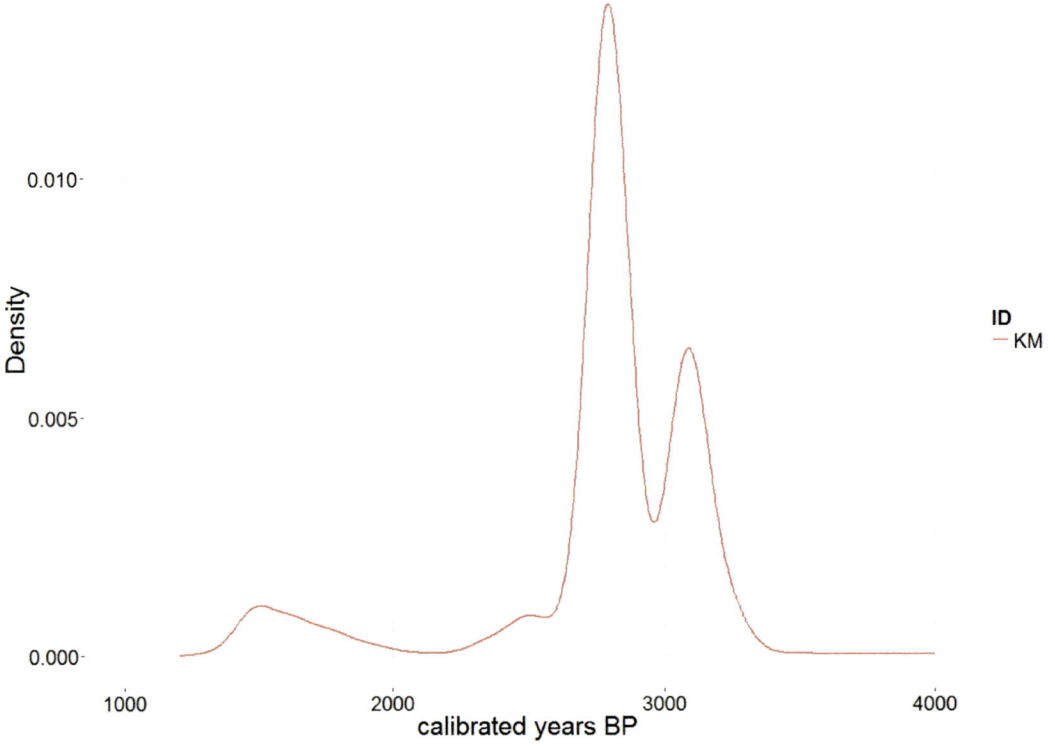

Figure 5.1: The density distribution of radiocarbon the dates from the Kimpo-Yangchon site, using the R package BChron (the dates were calibrated using the "intcal13" calibration curve, cf. Reimer et al., 2013)

5.1.1. Sampling

5.1.1.1. Organic geochemical analysis

At least two samples were collected from each of the houses, except those which did not yield pottery, and of which the date could not be estimated. If available, three samples were collected from one house. One sample was collected from some house pits which did not yield enough potsherds. Researches have showed that the pottery for the ordinary day-to-day subsistence around this period tend to have rather monotonous characteristics in terms of shape and size (Bae, 2007; Shoda, 2008). Therefore, the shape and size of the pottery were relatively not critical issues for sampling. According to the experimental analysis of Evershed (2008a), the rim and upper body parts of pots are where organic residues are the most concentrated after cooking (cf. A. Barker et al., 2012; Eerkens, 2007). Ethnographic observations showed that generally, high-temperature boiling is regarded as a particularly effective cooking method in the preparation of faunal and floral resources in pots (Crown & Wills, 1995; Stahl, 1989; Wandsnider, 1997). During this process, convection currents of boiling water push extracted lipids from food stuffs to the pot wall. Since lipids float on water, they tend to accumulate and penetrate into the wall of the upper body and rim of the pot. Taking these facts as criteria, a total of 49 samples were collected (Table 5.2, Figure 5.3). Since some of the house pits were dated by the AMS radiocarbon dating, if there are available dates from the house pits where the samples were collected, I indicated them in Table 5.2.

Figure 5.2: The location of the four sites analyzed in this study

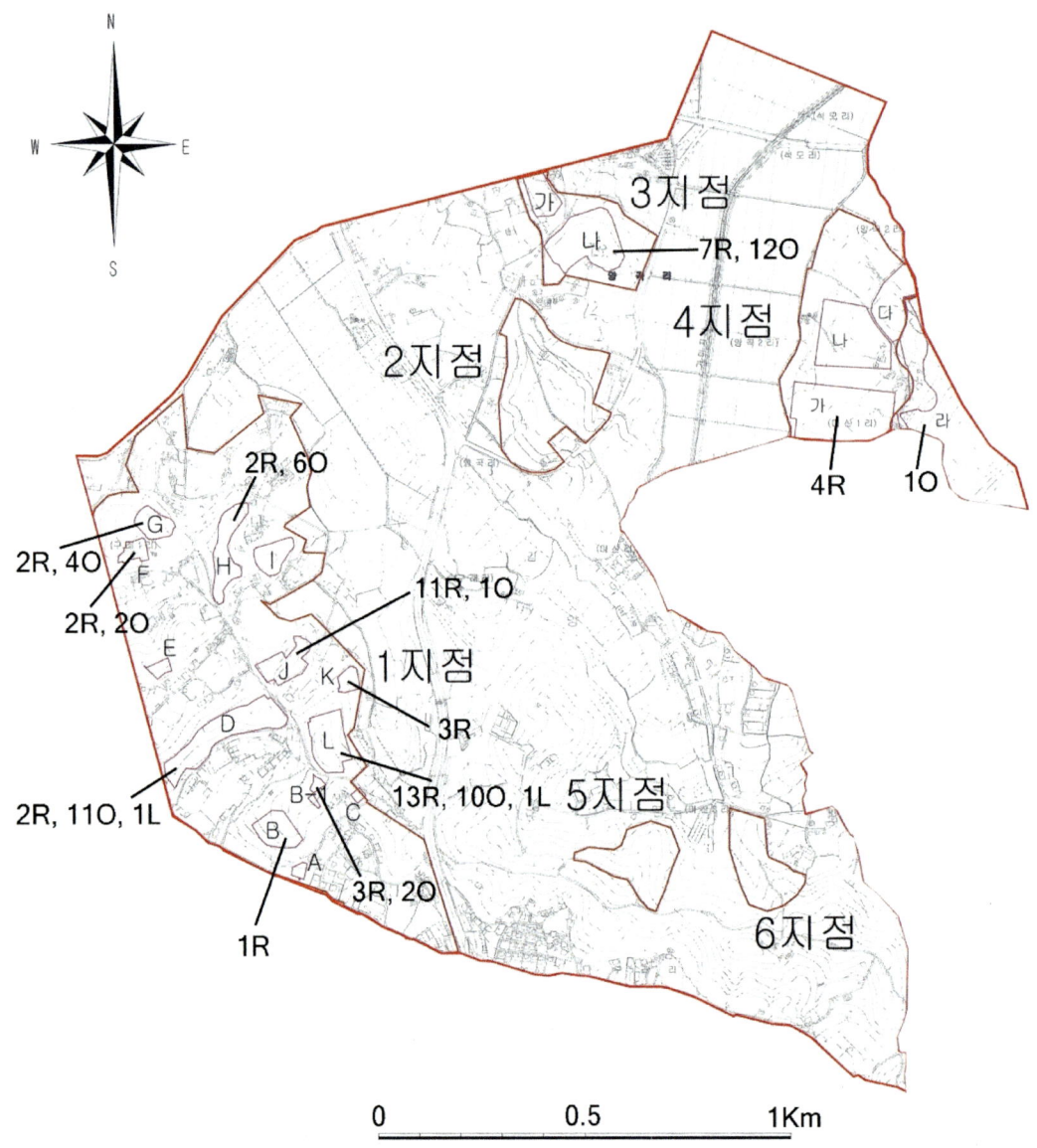

Figure 5.3: The site plan of the Kimpo-Yangchon site and the location/number of the samples taken for the radiocarbon dating (R), organic geochemical analysis (O), and luminescence dating (L) (B. M. Kim et al., 2013)

Figure 5.4: Some of the artifacts uncovered during the excavation of the Kimpo-Yangchon site: semilunar shaped knife (upper-left), pots (right; rim-punctuation: upper-right), and arrowheads (down-left)

Sample No.	Location/house pit No.	Part	C^{14} date (BP; uncalibrated)
KIM030	Area 2-3 "Na"/No.3	Body	2660±50
KIM031	Area 2-3 "Na"/No.3	Body	
KIM032	Area 2-3 "Na"/No.7	Body	2710±50
KIM033	Area 2-3 "Na"/No.7	Body	2710±50
KIM034	Area 2-3 "Na"/No.7	Body	2760±50
KIM035	Area 2-3 "Na"/No.8	Body	2920±50
KIM036	Area 2-3 "Na"/No.8	Body	2920±50
KIM037	Area 2-3 "Na"/No.8	Body	2920±50
KIM038	Area 2-3 "Na"/No.11	Body	
KIM039	Area 2-1 "L"/No.3	Body	2960±50
KIM040	Area 2-1 "L"/No.3	Body	2960±50
KIM041	Area 2-1 "L"/No.3	Body	2960±50
KIM042	Area 2-1 "L"/No.10	Rim	2820±50
KIM043	Area 2-1 "L"/No.10	Body	2820±50
KIM044	Area 2-1 "L"/No.11	Body	2910±50
KIM045	Area 2-1 "L"/No.11	Body	2910±50
KIM046	Area 2-1 "F"/No.1	Body	4530±50 (Chulmun)
KIM047	Area 2-1 "F"/No.1	Body	4530±50 (Chulmun)
KIM048	Area 2-1 "B-1"/No.1	Body	
KIM049	Area 2-1 "D"/No.14	Body	
KIM050	Area 2-1 "D"/No.14	Body	
KIM051	Area 2-1 "D"/No.8	Body	
KIM052	Area 2-1 "D"/No.8	Body	
KIM053	Area 2-1 "D"/No.9	Body	
KIM054	Area 2-1 "D"/No.9	Body	
KIM055	Area 2-1 "D"/No.15	Body	
KIM056	Area 2-1 "D"/No.15	Body	
KIM057	Area 2-1 "L"/No.3	Body	
KIM058	Area 2-1 "D"/No.10	Body	
KIM059	Area 2-3 "NA"/No.5	Body	
KIM060	Area 2-3 "NA"/No.5	Body	
KIM061	Area 2-1 "G"/No.3	Body	
KIM062	Area 2-1 "G"/No.3	Body	
KIM063	Area 2-1 "H"/No.5	Body	2380±40
KIM064	Area 2-1 "H"/No.5	Body	2380±40
KIM065	Area 2-1 "H"/No.12	Body	2770±40
KIM066	Area 2-1 "H"/No.12	Body	2770±40
KIM067	Area 2-1 "H"/No.20	Body	
KIM068	Area 2-1 "H"/No.20	Body	
KIM069	Area 2-4 "Ra"/No.20	Body	
KIM070	Area 2-3 "Na"/No.3	Body	
KIM071	Area 2-1 "B-1"/No.3	Body	
KIM072	Area 2-1 "D"/No.14	Body	
KIM073	Area 2-1 "G"/No.5	Rim	
KIM074	Area 2-1 "G"/No.5	Body	
KIM075	Area 2-1 "J"/No.1	Body	
KIM076	Area 2-1 "L"/No.1	Body	
KIM077	Area 2-1 "D"/No.9	Body	
KIM078	Area 2-1 "L"/No.9	Rim	

Table 5.2: The samples collected from the Kimpo-Yangchon site for the organic geochemical analysis

5.1.1.2. Luminescence dating

For the luminescence dating two samples were collected. Both of the samples were collected from a house which was not dated (Table 5.3, Figure 5.3).

Sample No.	Location/house pit No.	Part	Depth (m)
U3045	Area 2-1 "L"/No.3	Body	0.3
U3046	Area 2-1 "D"/No.10	Body	0.3

Table 5.3: The samples collected from the Kimpo-Yangchon site for the luminescence dating

5.1.2. Organic geochemical results

Before collecting 49 samples from the Kimpo-Yangchon site for the organic geochemical analysis in this study, 25 samples were collected for a preliminary analysis. They were all collected based on the same criteria that were mentioned in the "sampling" section. The purpose of the preliminary analysis is to ascertain the applicability of the organic geochemical analysis to examining the potteries from the central part of the Korean Peninsula. The samples were analyzed in accordance with the well-known standard solvent extraction protocol that demands the use of solvent mixture (chloroform-methanol 2 : 1 v/v; cf. chapter four), at the organic geochemistry unit, University of Bristol, under the guidance of Dr. Richard P. Evershed. Unfortunately, since the lipid concentration of the samples was so low, I was not able to extract an analyzable amount of lipids from those 25 samples (cf. Figure 4.8a). Following Dr. Evershed's suggestion the direction of examination was changed to employ the methanolic acid extraction protocol (Correa- Ascencio & Evershed, 2014, cf. chapter four). In this study, all the 49 samples from the Kimpo-Yangchon site were analyzed by the acid extraction protocol.

Figure 5.5, 5.6, 5.7, and 5.8 show the results of the organic geochemical analyses. Among the 49 samples, I was able to analyze 20. 29 samples had to be omitted mainly due to contamination and low concentration of lipids. In spite of going through the cleaning process of samples using drill bits to minimize contamination, in accordance with the standard protocol (cf. Chapter four), not all the sherds were suitable for the analysis. This is mainly because of poor handling of the pottery during the excavation and curation processes. Generally, the most frequently observed compounds in archaeological lipid residues are palmitic (C16:0) and stearic (C18:0) fatty acids (Evershed, 2008a). As expected, the organic compounds of all samples were dominated by those two saturated fatty acids (Figure 5.5). This means those organic compounds were highly degraded in soil during several thousand years of post-depositional processes (cf. Chapter four). Nevertheless, with the results of GC-MS analyses, I was able to identify both major short- and long-chain saturated fatty acids including C14:0, C15:0, C15:1, C17:0, C20:0, C22:0, and C24:0. Generally, a high C18:0 saturated alkanoic acid content indicates an animal source (Enser, 1991; Copley et al., 2005c; Evershed et al., 2002: 664). Most of the lipid residues analyzed showed a high concentration of C18:0 fatty acid (Figure 5.5).

There are compounds which are only found in certain food groups. Especially, phytanic acid (3,7,11,15-tetramethylhexadecanoic acid) and 4,8,12-TMTD (4,8,12-trimethyltridecanoic acid) are isoprenoid compounds which are mostly found in a particularly high concentration in marine animals (Evershed, 2008b, cf. Chapter four). Along with thermally produced long-chain ω-(o-alkylphenyl)alkanoic acids, these compounds are indicators of aquatic/marine resources (Craig et al., 2011; Evershed et al., 2008). Since the Kimpo-Yangchon site is only 4 kilometers apart from the Han river (Figure 5.2), it is essential to know whether its dwellers relied on aquatic resources. Among those 20 samples, one sample showed the presence of phytanic acid (KIM061), no other aquatic biomarker was detected (cf. Figure 5.5). Since phytanic acid is also found in the tissues of ruminant animals, I conclude phytanic acid alone from one sample is not a sufficient indicator of aquatic resource on its own (cf. Heron and Craig, 2015). However, I cannot exclude a possibility of fish as a part of diet at the Kimpo-Yangchon site, as it could have been cooked in a direct fire, leaving no signal on pottery.

The results of the isotope analyses (Figure 5.6; 5.7; 5.8) effected on palmitic (C16:0) and stearic (C18:0) fatty acids on the samples show more interesting characteristics of these ancient farmers' diet. They indicate that they consumed various food stuffs including pork, ruminants, and aquatic resources (Fresh water and Marine). Many samples indicate that the pots from which they came were used for processing multiple foodstuffs. The dominant food classes were the pork and the aquatic resources.

Though some of the samples showed the presence of C_3 plant oils, it has to be carefully considered whether this indicates rice. Firstly, C_3 plants include not only rice, but also legumes and barley. As G. Lee (2011) mentioned, we have pollen data from 5,500 BP to 2,600 BP showing the ancient farmers of the Korean Peninsula utilized soybean (Glycine max) and azuki (Vignaaugularis) as subsistence resources. Therefore, it is impetuous to argue that the detected C_3 plant oils are from rice alone. Secondly, since the area of C_3 plant oils in Figure 5.6 could indicate the mixture of pork and ruminant adipose (cf. Chapter four), we do not have any assurance that the C_3 plant oils of which the presence is indicated by those five samples are actually plant oil. Lastly, As I mentioned above, Most of the lipid residues from Kimpo-Yanchon showed a high concentration of C18:0 fatty acid, which generally indicates an animal source. However, this does not indicate harvested corps were never processed or cooked in a pot at the Kimpo-Yangchon site. Animal (and fish) fats are much more concentrated than plant oil in general and the former often mask the signal of the latter, if they are cooked in a same pot. Therefore, I am not removing the possibility that some (or most) of the pots were used for cooking both animals and grains.

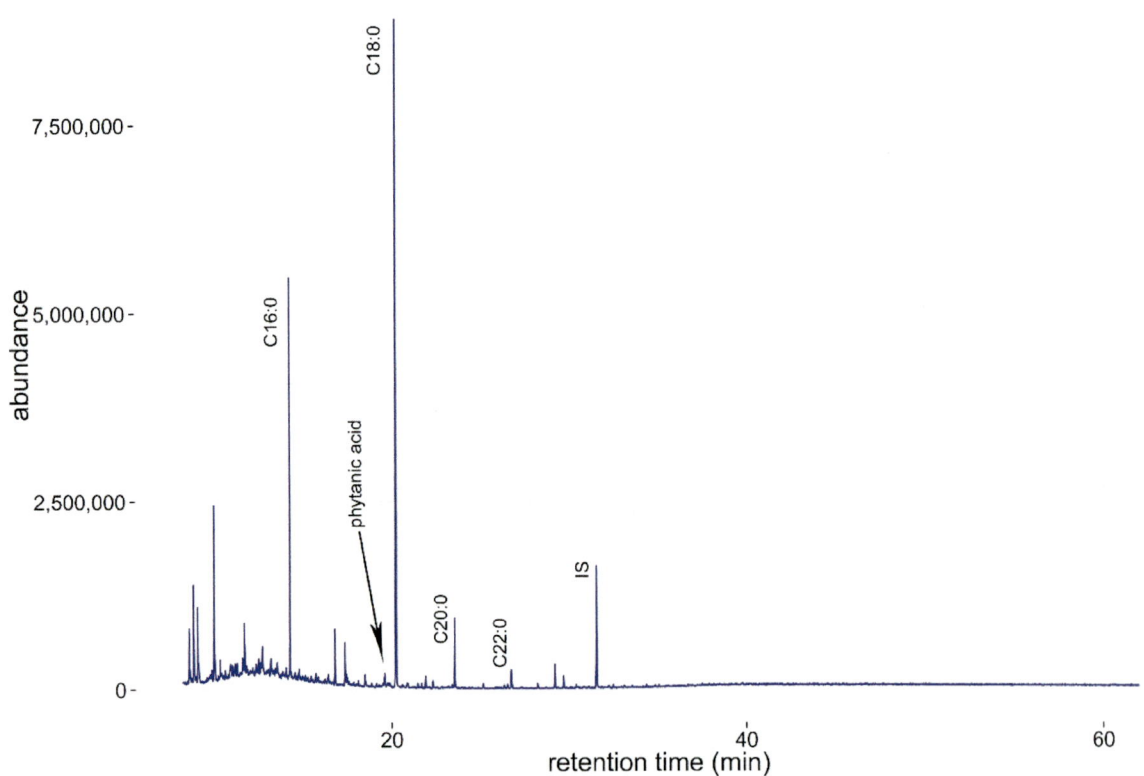

Figure 5.5: The result chromatogram of the GC-MS analysis of one of the samples from the Kimpo-Yangchon site (KIM061), using R version 3.2.0. Due to degradation, we usually observe medium- and long-chain saturated fatty acids. 5-α Cholestane was added as an internal standard (IS = 132 ng / microliter)

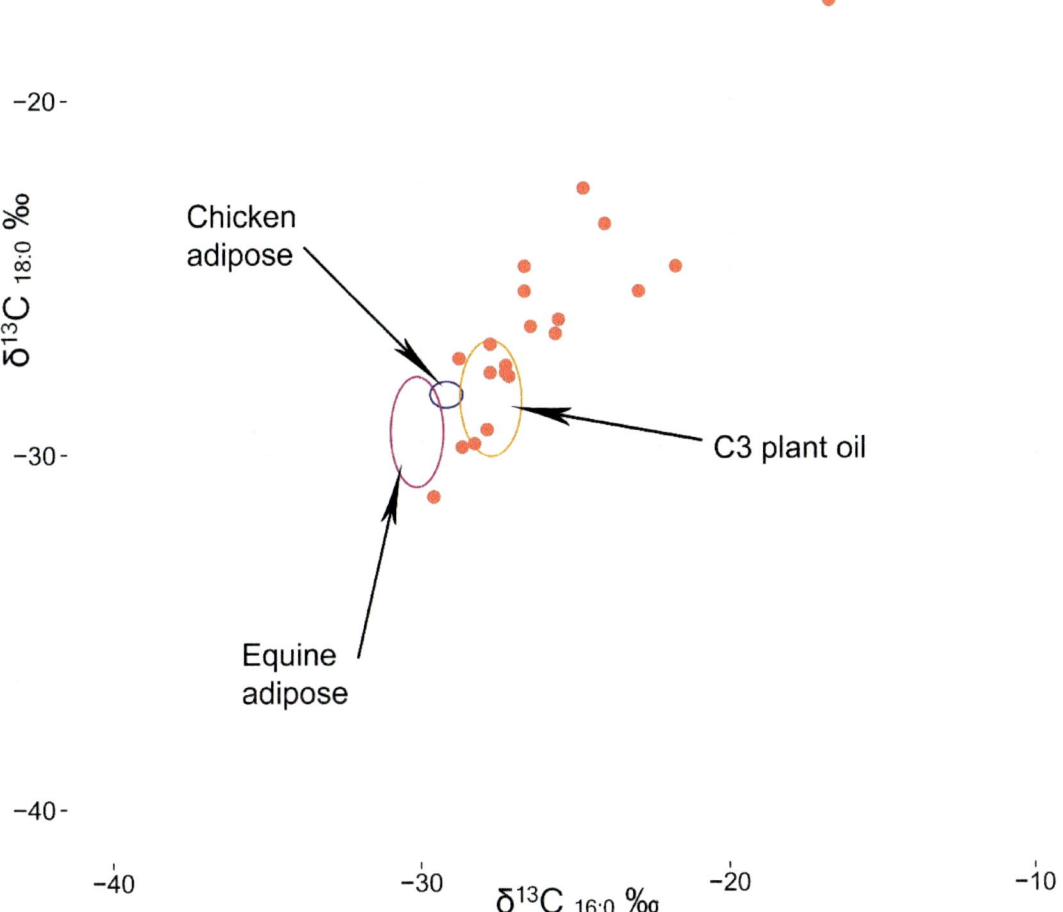

Figure 5.6: The results of CSIA by GC-C-IRMS of the samples from the Kimpo-Yangchon site using the available references (cf. Dudd & Evershed, 1998; Dudd et al., 1999; Steele et al., 2010)

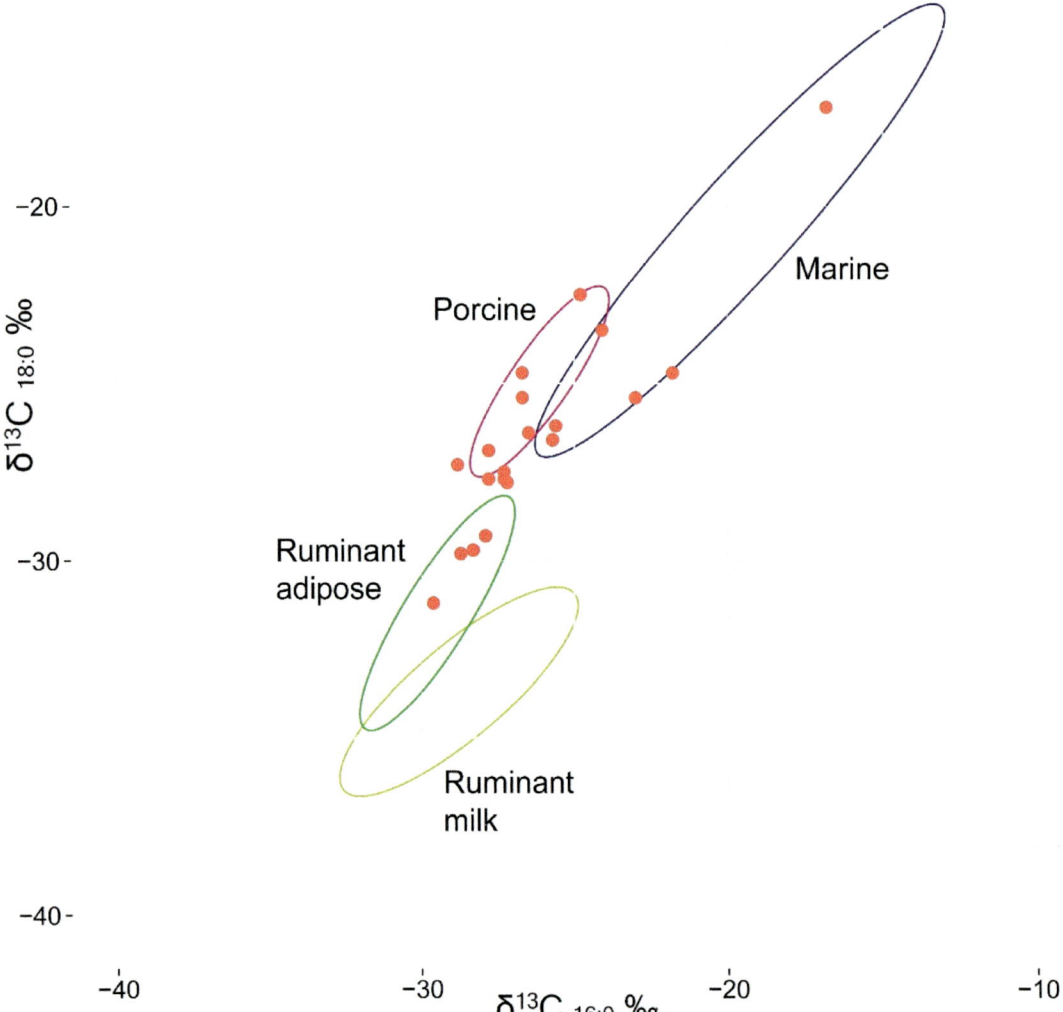

Figure 5.7: The results of CSIA by GC-C-IRMS of the samples from the Kimpo-Yangchon site using the reference from Craig et al. (2011)

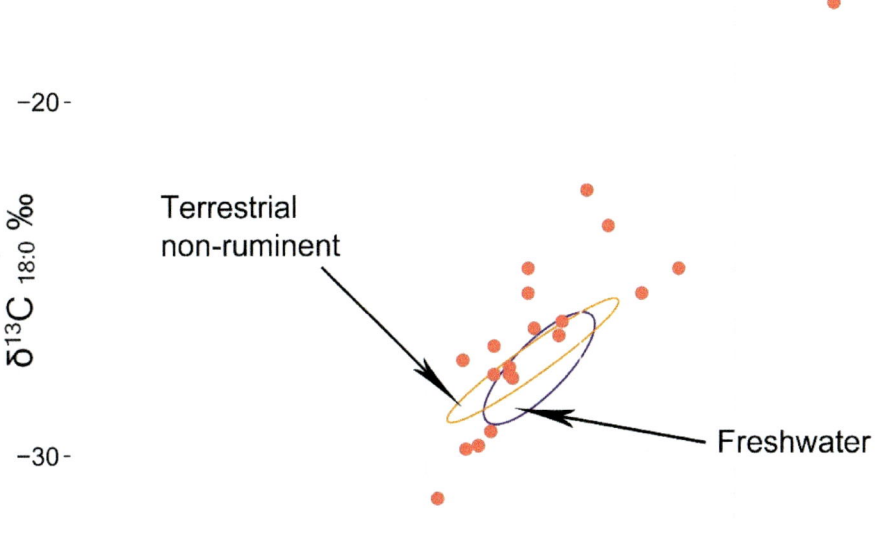

Figure 5.8: The results of CSIA by GC-C-IRMS of the samples from the Kimpo-Yangchon site using the reference from Craig et al. (2013)

5.1.3. Luminescence dating results

The samples were dated using TL, OSL, and IRSL at the luminescence dating lab, University of Washington. Table 5.4 shows the results of the luminescence dating. The OSL and TL ages were in agreement for the sample UW3045, and TL fading was not significant. The IRSL age is younger, probably due to the anomalous fading. The OSL age was the best estimate for the sample UW3046. The IRSL age for UW3056 was younger, probably due to the fading. Overall, the dates match with the main occupation period of the Kimpo-Yangchon site estimated by the radiocarbon dates (Figure 5.1).

Lab. No	Depth (m)	Water Content (%)	Dose rate* (Gy/ka)	De (Gy)			Age
				TL	OSL	IRSL	
U3045	0.30	10.7	5.227±0.486	13.625±3.55	12.819±0.296	11.141±0.429	740±160 BC
U3046	0.30	15.1	6.880±0.579	10.792±2.031	11.775±0.389	10.16±0.445	740±180 BC (OSL)

Table 5.4: The results of the luminescence dating of the potsherd samples from the Kimpo-Yangchon site.

5.2. Sosa-Dong

The Sosa-Dong site is located on the low hill of Sosa-Dong, Pyeongtaek city, Gyeonggi province. The site is about 2.5 kilometers north of the Anseong stream (Figure 2.7; 5.2). The excavation was conducted by Korea institute of Heritage, from September 2004 to September 2006 (B. M. Kim et al., 2008). The site includes various archaeological phenomena such as house pits, mound burials, pit graves, pit features and ditches which belong to different time periods from the Mumun period to the historical Joseon Dynasty (AD 1392 - 1897).

A total of 81 Mumun period house pits were found. Based on the results of the radiocarbon dating of charcoal from the house pits (B. M. Kim et al., 2008, Table 5.5), it is inferred that the site goes back to the times as early as the incipient/early stage of the Mumun period, or as late as the middle/late Mumun period (cf. Figure 5.11; 6.3). The house pits are classified into four types based on their shape: square, circular, rectangular, and longhouse. The rectangular and longhouse pits were built around the early stage of the Mumun period (3000 - 2700 BP); and the square and circular pits near the late Mumun period (2500 - 2300 BP). According to the radiocarbon dating on the charcoal from hearths in the house pits, the site has a chronological void from 2700 BP to 2500 BP (Table 5.5). Some of these houses incorporate hearths, storage pits and ditches as interior features. Most of the potteries have the rim-punctuation or a combination of lip-scoring/rim-punctuation; and others a combination of double-rim/short slanted incision or rim-punctuation/short slanted incision (Figure 5.9; 2.4). As for the ground stone tools, arrowheads, daggers, chisels and axes were found (Figure 5.9). As for the farming tools, semi-lunar shaped stone knives (Figure 2.6b) and mortars/pestles were found. Especially, carbonized 46 rice (*Oryza sativa*; Figure 5.10a) and 31 possible barley (*Hodeum vulgare* L.; Figure 5.10b) grains were found inside of one house pit, near the hearth (Area "Ga"/No. 10).

The overall archaeological phenomena of the Sosa-Dong site are quite similar to those of the Kimpo-Yangchon site. The composition of different types of house pits, potteries and stone artifacts clearly indicate the resemblance between the two sites. Probably one of the most interesting features of the Sosa-Dong site compared with the Kimpo-Yangchon site is carbonized rice and possible barley grains. Considering their 'burnt' condition, it is beyond all doubt that rice and barley were cooked for consumption.

Location/house pit No.	Cultural historical period	C^{14} date (BP; uncalibrated)	Calendar date
Area "La"/No. 20	Mumun	3010±60	BC 1240
Area "Da"/No. 5	Mumun	2990±50	BC 1220
Area "Da"/No. 6	Mumun	2990±50	BC 1220
Area "Ga"/No. 17	Mumun	2950±50	BC 1160
Area "Ga"/No. 7	Mumun	2930±60	BC 1150
Area "Da"/No. 7	Mumun	2930±50	BC 1150
Area "La"/No. 10	Mumun	2900±50	BC 1120
Area "Ga"/No. 2	Mumun	2850±60	BC 1060
Area "Ga"/No. 10	Mumun	2840±50	BC 1050
Area "Ga"/No. 14	Mumun	2850±50	BC 1050
Area "Ga"/No. 16	Mumun	2840±50	BC 1050
Area "Ga"/No. 18	Mumun	2840±50	BC 1050
Area "Ga"/No. 28	Mumun	2850±50	BC 1050
Area "Da"/No. 4	Mumun	2810±50	BC 980
Area "Ga"/No. 20	Mumun	2750±50	BC 910
Area "La"/No. 4	Mumun	2740±50	BC 900
Chronological void			
Area "Ga"/No. 13	Mumun	2550±50	BC 670
Area "La"/No. 7	Mumun	2470±80	BC 600
Area "Ga"/No. 15	Mumun	2470±60	BC 590
Area "Ga"/No. 4	Mumun	2300±50	BC 310

Table 5.5: The results of AMS radiocarbon dating of the Sosa-Dong site

Figure 5.9: Some of the artifacts uncovered during the excavation of the Sosa-Dong site including potsherd, arrowheads and stone chisel. The potsherd in the picture has the rim-punctuation/short slanted incision.

Figure 5.10: (a): The carbonized rice grains (*Oryza sativa*) and (b): possible barley (*Hodeum vulgare* L.) grains excavated in the Area "Ga" house pit No. 10 (B. M. Kim et al, 2008)

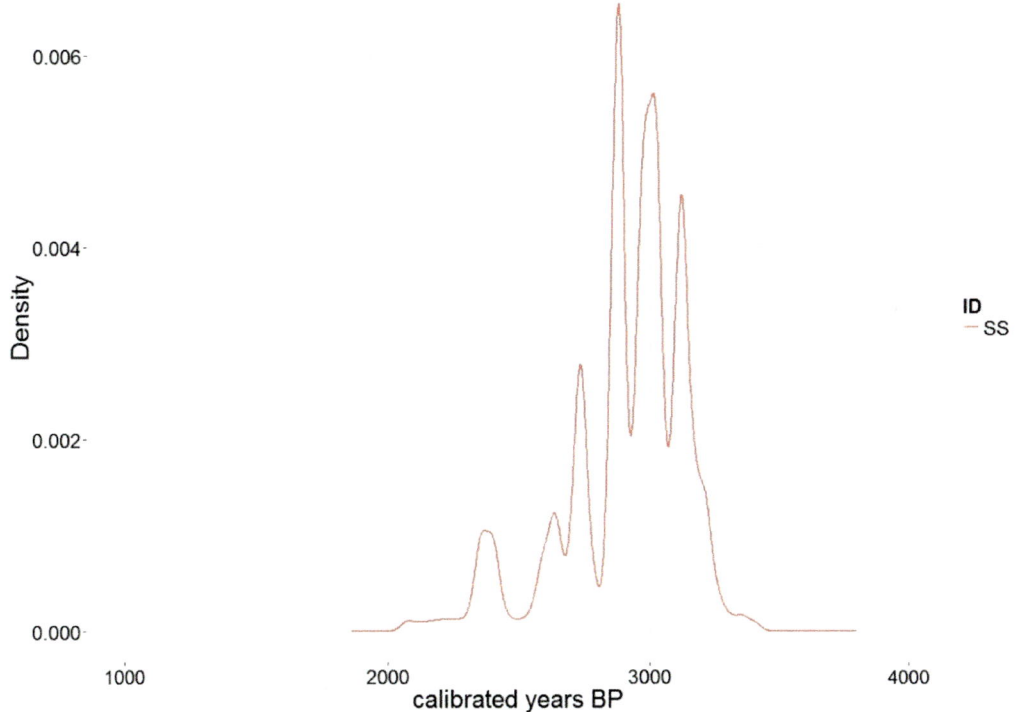

Figure 5.11: The density distribution of the radiocarbon dates from the Sosa-Dong site, using the R package BChron (the dates were calibrated using the "intcal13" calibration curve, cf. Reimer et al., 2013)

5.2.1. Sampling

5.2.1.1. Organic geochemical analysis

The general sampling strategy for the organic geochemical analysis on the Sosa-Dong site is quite similar to that on the Kimpo-Yangchon site. At least two samples were collected from each of the houses, except those which did not yield pottery, and whose date could not be estimated. If available, three samples were collected from one house. One sample was collected from some house pits which did not yield enough potsherds. The shape and size of the pots were not considered, for the pottery for the ordinary day-to-day subsistence around this period tend to have rather monotonous characteristics in terms of shape and size (Bae, 2007; Shoda, 2008). Following the criteria of Evershed (2008a, Figure 5.12), the rim and upper body parts were chosen and a total of 37 samples were collected (Table 5.6, Figure 5.13). If there are available radiocarbon dates from the house pits where the samples were collected, I indicated them in Table 5.6.

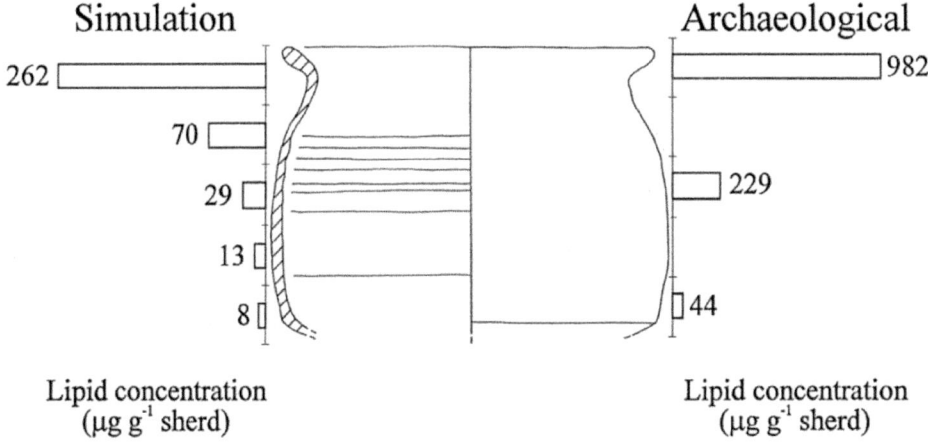

Figure 5.12: Diagram showing the lipid concentration of each body part from the both experimental and archaeological sherd samples (adapted from Evershed, 2008a: 32)

Sample No.	Location/house pit No.	Part	C^{14} date (BP; uncalibrated)
SOS030	Area "La"/No. 14	Body	
SOS031	Area "La"/No. 4	Body	
SOS032	Area "La"/No. 4	Rim	
SOS033	Area "La"/No. 4	Rim	
SOS034	Area "Ga"/No. 7	Body	2930±60
SOS035	Area "Ga"/No. 10	Body	2840±50
SOS036	Area "Ga"/No. 10	Body	2840±50
SOS037	Area "Ga"/No. 14	Body	2850±50
SOS038	Area "Ga"/No. 14	Body	2850±50
SOS039	Area "La"/No. 11	Rim	
SOS040	Area "La"/No. 11	Body	
SOS041	Area "La"/No. 11	Body	
SOS042	Area "Ga"/No. 23	Body	
SOS043	Area "Ga"/No. 23	Body	
SOS044	Area "Ga"/No. 24	Body	
SOS045	Area "Ga"/No. 24	Body	
SOS046	Area "Ga"/No. 25	Body	
SOS047	Area "Ga"/No. 25	Body	

Sample No.	Location/house pit No.	Part	
SOS048	Area "La"/No. 15	Rim	
SOS049	Area "La"/No. 15	Rim	
SOS050	Area "La"/No. 15	Rim	
SOS051	Area "La"/No. 2	Body	
SOS052	Area "La"/No. 2	Body	
SOS053	Area "La"/No. 5	Body	
SOS054	Area "La"/No. 5	Body	
SOS055	Area "La"/No. 10	Body	2900±50
SOS056	Area "La"/No. 10	Rim	2900±50
SOS057	Area "La"/No. 19	Body	
SOS058	Area "La"/No. 19	Body	
SOS059	Area "La"/No. 18	Body	
SOS060	Area "La"/No. 18	Body	
SOS061	Area "La"/No. 31	Body	
SOS062	Area "La"/No. 31	Body	
SOS063	Area "La"/No. 31	Body	
SOS064	Area "La"/No. 32	Body	
SOS065	Area "La"/No. 32	Body	
SOS066	Area "La"/No. 36	Body	

Table 5.6. The samples collected from the Sosa-Dong site for the organic geochemical analyses

5.2.1.2. Luminescence dating

As at the Kimpo-Yangchon site, two samples were collected for the luminescence dating. One of the two samples was collected from a house which had been dated by the radiocarbon dating, and the other from another which had not been (Table 5.7, Figure 5.13).

Sample No.	Location/house pit No.	Part	Depth (m)
U3042	Area "La"/No.4	Body	0.3
U3043	Area "La"/No.14	Body	0.3

Table 5.7: The samples collected from the Sosa-Dong site for the luminescence dating

5.2.2. Organic geochemical results

As at the Kimpo-Yangchon site, before collecting 37 samples, 21 samples were collected for a preliminary analysis to ensure the analytical protocol. The samples were collected based on the same sampling strategy in this study and analyzed by the standard solvent extraction protocol (chloroform-methanol 2 : 1 v/v; cf. chapter four) at the organic geochemistry unit, University of Bristol. However, it was nearly impossible to extract lipids from those samples, due to their low concentration (cf. Figure 4.8a). Under this circumstance, the direction of examination was changed to employ the methanolic acid extraction protocol (Correa-Ascencio & Evershed, 2014, cf. chapter four). In this study, all the 37 samples from the Sosa-Dong site were analyzed by the acid extraction protocol.

Figure 5.13: The site plan of the Sosa-Dong site and the location of the samples taken for the radiocarbon dating (R), organic geochemical analysis (O), and luminescence dating (L) (B. M. Kim et al., 2008)

Figure 5.14, 5.15, 5.16, and 5.17 show the results of the organic geochemical analyses. Among the 37 samples, 28 were analyzable. Nine samples had to be omitted mainly due to contamination and the low concentration of lipids. Compared with that of the Kimpo-Yangchon site (20 analyzable samples among 49), this recovery rate is quite high. Considering that there are spatio-temporal similarities between the two sites, their difference in recovery rate of samples probably means the potsherds were more carefully treated during the excavation and curation processes in case of the Sosa-Dong site.

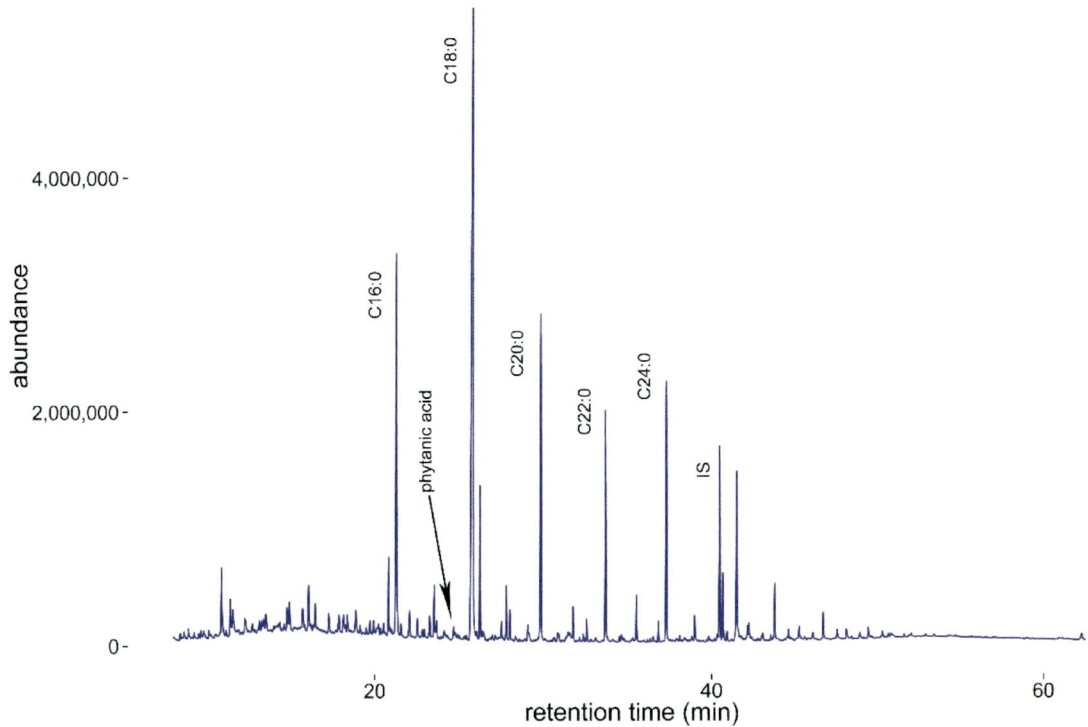

Figure 5.14: The result chromatogram of the GC-MS analysis of one of the samples from the Sosa-Dong site (SOS049), using R version 3.2.0. Due to degradation, we usually observe medium- and long-chain saturated fatty acids. 5-α Cholestane was added as an internal standard (IS = 132 ng / microliter)

As I mentioned above, the most frequently observed compounds in archaeological lipid residues are the palmitic (C16:0) and stearic (C18:0) fatty acids (Evershed, 2008a). The Sosa-Dong site was not an exception, and the organic compounds of all samples were dominated by those two saturated fatty acids, due to the degradation in soil during several thousand years of post-depositional processes (Figure 5.14). Along with the C16:0 and C18:0 fatty acids, I was able to identify both major short- and long-chain saturated fatty acids including C13:0, C14:0, C15:0, C17:0, C18:0, C20:0, C22:0, C23:0, and C24:0. Most of the lipid residues analyzed showed a high concentration of C18:0 fatty acid (Figure 5.14), which generally indicates an animal source (Enser 1991; Copley et al. 2005c; Evershed et al. 2002: p. 664).

The geographic location of the Sosa-Dong site is quite similarly to that of the Kimpo-Yangchon site. The site is only 2.5 kilometers apart from the Anseong stream, and also close to the Yellow Sea (Figure 5.2). This means it is quite possible that the farmers of the Sosa-Dong site performed fishing also. During the excavation of the Sosa-Dong site, a total of 17 net sinkers were found. In this regard, it is essential to know whether the dwellers of the Sosa-Dong site relied on aquatic resources. According to Evershed et al. (2008), phytanic acid (3,7,11,15-tetramethylhexadecanoic acid), 4,8,12-TMTD (4,8,12- trimethyltridecanoic acid) and thermally produced long-chain ω-(o-alkylphenyl)alkanoic acids are the indicators of aquatic/marine resources (cf. Craig et al., 2011). Among those 28 samples, two samples showed the presence of phytanic acid (SOS049, SOS056). As I have mentioned above, phytanic acid is also found in the tissues of ruminant animals. Therefore, phytanic acid alone is not a sufficient indicator of aquatic resource on its own (cf. Heron and Craig, 2015). However, I cannot exclude a possibility that of fish could have been cooked in a direct fire.

The results of the isotope analysis effected on palmitic (C16:0) and stearic (C18:0) fatty acids on the samples show a varied diet of these ancient farmers. The results (Figure 5.15; 5.16; 5.17) indicate that they consumed several food stuffs including pork, aquatic resources. The diet of the ancient dwellers of the Sosa-Dong site was dominated by pork and aquatic resources. As a whole, the diet pattern of the Sosa-Dong site is somewhat similar to that of the Kimpo-Yangchon site.

5.2.3. Luminescence dating results

The samples were dated using TL, OSL, and IRSL at the luminescence dating lab, University of Washington. Table 5.8 shows the results of the luminescence dating. The OSL and TL ages were in agreement for the sample UW3042, and TL fading was not significant. The IRSL age was younger, probably because of the fading of feldspar. The OSL, IRSL, and TL ages were in agreement for the sample UW3043 (the fading was not significant). The dates were slightly younger than the main occupation period of the Sosa-Dong site estimated by the radiocarbon dates (Figure 5.11).

Lab. No	Depth (m)	Water Content (%)	Dose rate* (Gy/ka)	De (Gy)			Age
				TL	OSL	IRSL	
U3042	0.30	18.4	7.872±0.475	20.97±1.59	14.487±0.43	13.194±0.307	650±140 BC
U3043	0.30	19.7	6.664±0.400	13.999±1.469	11.958±0.229	14.18±0.591	390±110 BC

Table 5.8: The results of the luminescence dating of the potsherd samples from the Sosa-Dong site

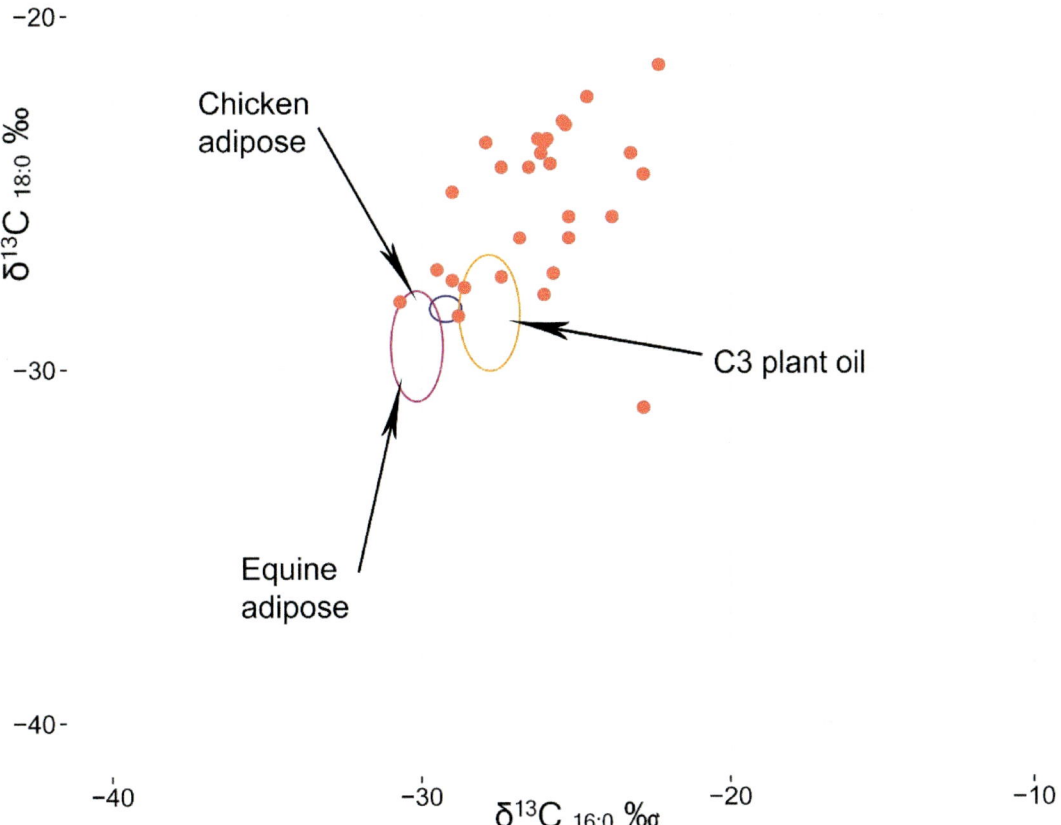

Figure 5.15: The results of CSIA by GC-C-IRMS of the samples from the Sosa-Dong site using the available references (cf. Dudd & Evershed, 1998; Dudd et al., 1999; Steele et al., 2010)

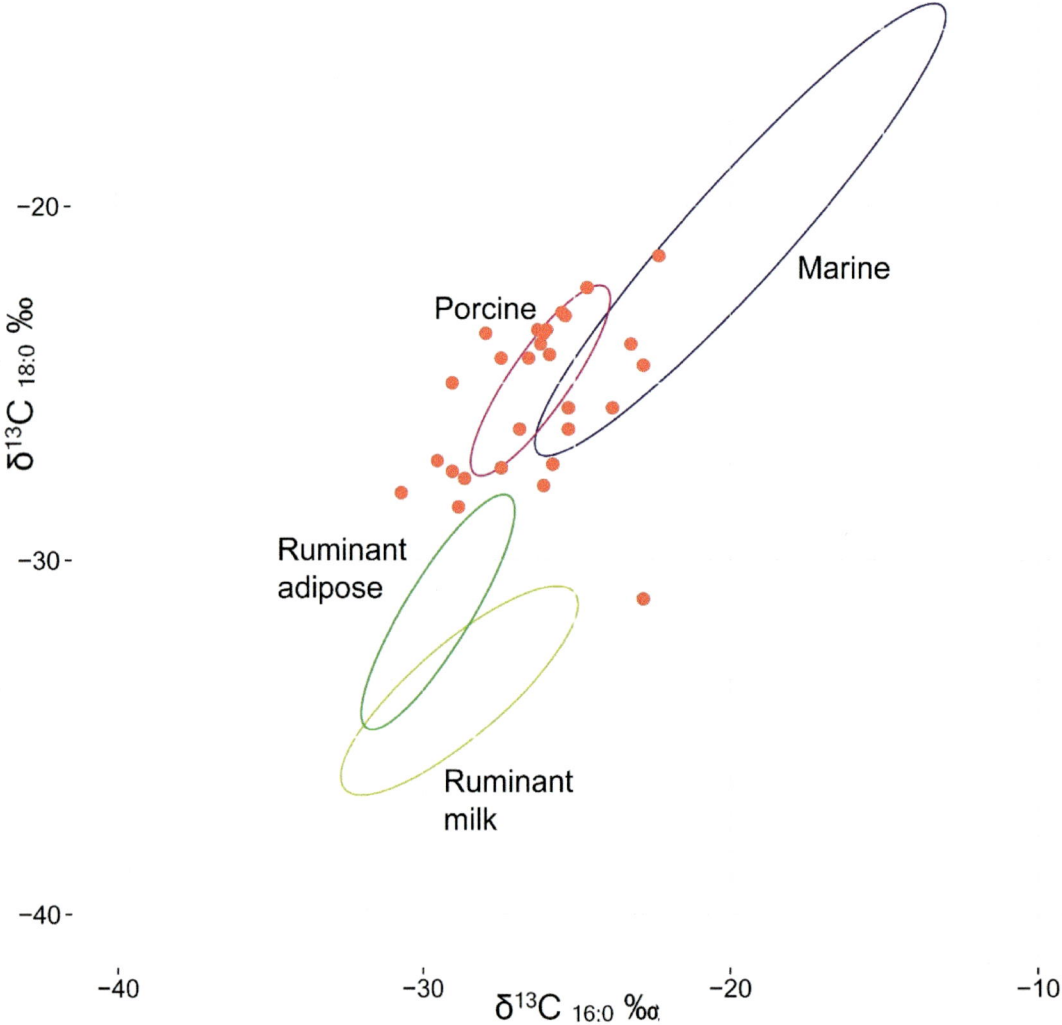

Figure 5.16: The results of CSIA by GC-C-IRMS of the samples from the Sosa-Dong site using the reference from Craig et al. (2011)

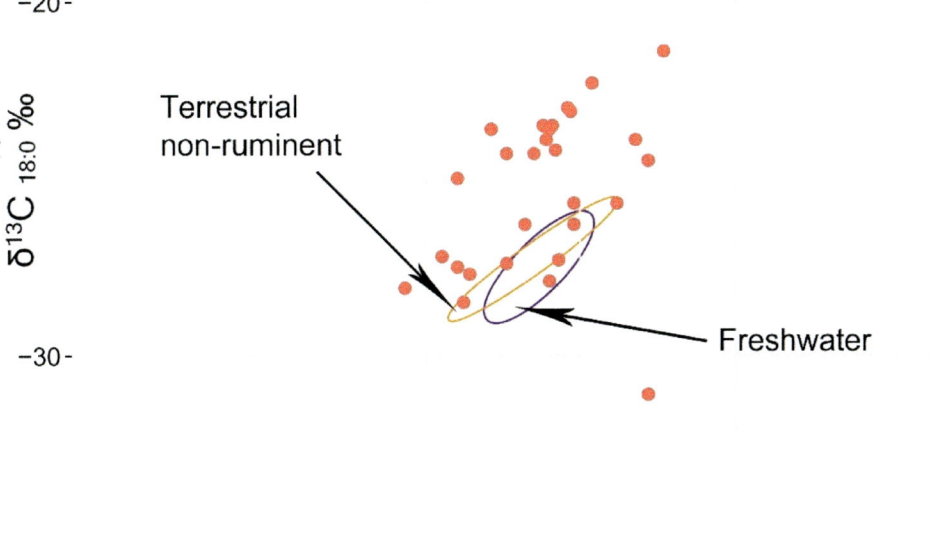

Figure 5.17: The results of CSIA by GC-C-IRMS of the samples from the Sosa-Dong site using the reference from Craig et al. (2013)

5.3. Songguk-Ri

Among the thousands of prehistoric archaeological phenomena in the Korean Peninsula, probably one of the most well-known and thoroughly studied sites is the Songguk-Ri site. Located in Buyeo city, Chungnam province, South Korea, it belongs to the Middle and Late Mumun period (Figure 2.7; 5.2). The initial excavation was conducted in 1975; and Songguk-Ri became the first archaeological site in Korea, which yielded bronze artifacts, tubular greenstone (jade) beads, typical un-patterned pottery and rounded pithouses with two post holes (Figure 5.18). These characteristic rounded pit houses were also found at other archaeological sites of later excavation, along with similar assemblages. It is why archaeologists recognized Songguk-Ri as a certain archaeological type of the Middle Mumun period, and designated both the formers and the latters 'the Songguk-ri Style'. Until now, the site has been excavated 14 times by different branches of the National Museum of Korea and the Korean National University of Cultural heritage (Buyeo National Museum, 2000; G. T. Kim et al., 2011; National Museum of Korea 1979, 1986, 1987).

Groups of pit-houses are found in various spots in an area of almost several square kilometers. The unpatterned potteries excavated from the site were named 'the Songguk-Ri style pottery (Figure 5.18a)'; and potteries of this style were found at many other sites in the central part of the Korean Peninsula with typical assemblages. The evidence of a wooden fence around the residential area indicates conflict and competition between the local Mumun societies (National Research Institute of Cultural Heritage, 2002). A number of smaller settlements presumed to be formed about the same period were found within the radius of several kilometers from Songguk-Ri. The site also includes stone-cist burials with a Liaoning-style bronze dagger, large tubular-shaped greenstone ornaments and a ground stone dagger (Figure 5.18). The high status materials (e.g. bronze dagger, green stone beads) in stone cist burials at the site and a number of small settlements around it led archaeologists to assume that in Songguk-Ri and its vicinity appeared the earliest form of social hierarchy in the ancient Korean Peninsula. With the importance of the site, it is registered as "Historical Site No. 249 (the Cultural heritage Administration of Korea)".

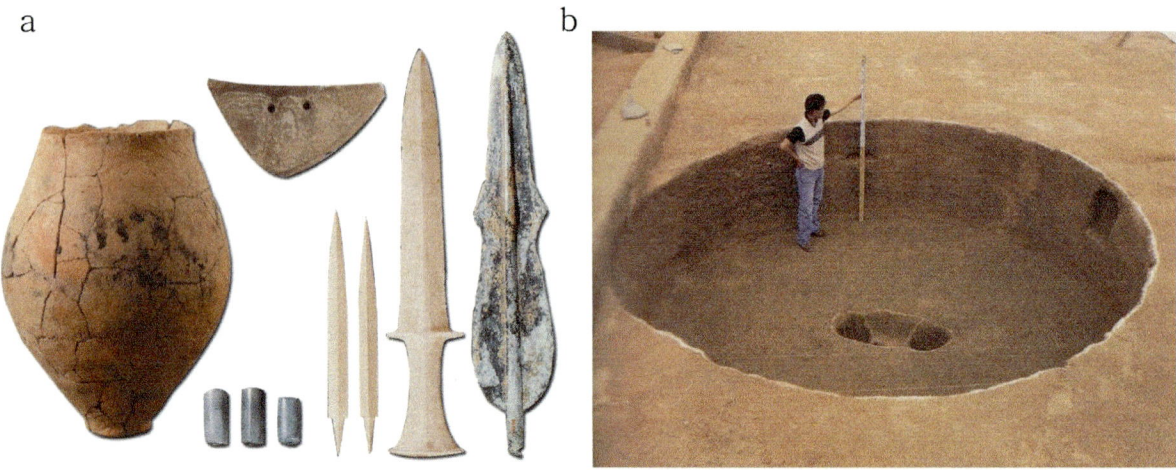

Figure 5.18: (a): some of the artifacts uncovered during the excavation of the Songguk-Ri site: pot, large tubular-shaped greenstone ornaments, semi-lunar shaped stone knife, arrowheads, ground stone dagger, and Liaoning-style bronze dagger (Yoon & Bae, 2010) (b): the "Songguk-Ri style" rounded pit-house with two post holes (Yoon & Bae, 2010)

The latest excavation of the Songguk-Ri site was conducted by the Korean National University of Cultural Heritage. The 12th to 14th excavations were held from April of 2008 to September of 2011 (G. T. Kim et al., 2011; 2013). As for the Mumun period, 47 house pits and 34 pit features were found. Based on the results of the radiocarbon dating of charcoal from the house pits and pit features (G. T. Kim et al., 2011, 2013, Table 5.9), the site was classified into the middle/late Mumun period (cf. Figure 5.19; 6.3). The house pits are classified into four types by their shape: circular, square, rectangular. No longhouse was found, for this type existed only during the incipient/early stage of the Mumun period. As for ground stone tools, arrowheads, semi-lunar shaped stone knives, spindle whorl, and pieces of green stone beads were excavated.

During the 14th excavation, several kinds of carbonized grains were found at 11 different features including house pits and pit features. The confirmed kinds were rice (Oryza sativa), foxtail millet (Setaria italica), broomcorn millet (Panicum Millaceum), soybean (Glycine max) and azuki (Vigna augularis). The two dominant grains were foxtail millet and rice, which occupied respectively about 65 and 32 percent of the identified ones, (their respective number: 5798 and 2892).

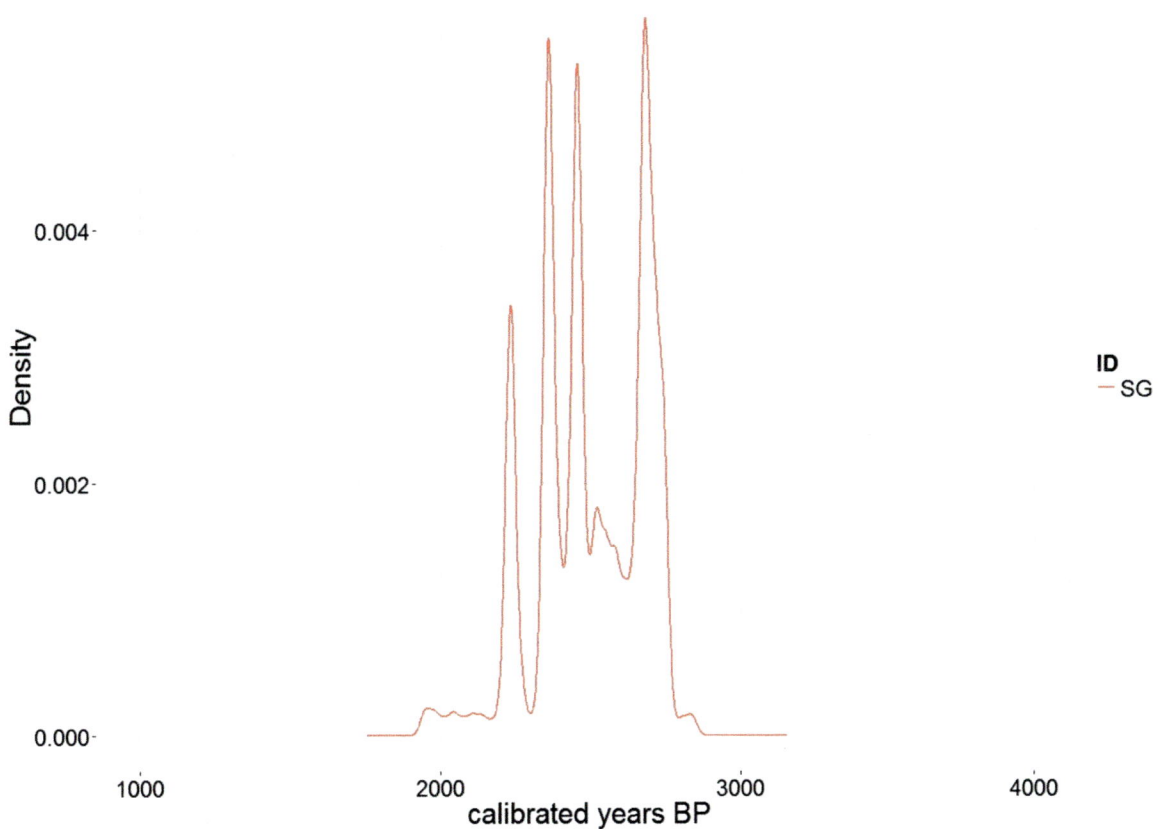

Figure 5.19: The density distribution of radiocarbon dates from the Songguk-Ri site, using the R package BChron (the dates were calibrated using the "intcal13" calibration curve, cf. Reimer et al., 2013)

house pit No.	Cultural historical period	C^{14} date (BP; uncalibrated)	Calendar date
No. 2	Mumun	2430±50	BC 475
No. 23	Mumun	2540±50	BC 660
No. 23	Mumun	2450±40	BC 580
No. 26	Mumun	2350±60	BC 450
No. 26	Mumun	2360±50	BC 450

No. 38	Mumun	2500±60	BC 655
No. 39	Mumun	2590±50	BC 785
No. 43	Mumun	2220±60	BC 260
No. 48	Mumun	2520±50	BC 595
No. 51	Mumun	2410±40	BC 470
No. 51	Mumun	2520±40	BC 650
No. 52	Mumun	2560±40	BC 680
No. 52	Mumun	2460±40	BC 580
No. 67	Mumun	2420±40	BC 470
No. 67	Mumun	2490±50	BC 650
No. 68	Mumun	2440±40	BC 580
No. 70	Mumun	2410±40	BC 470
No. 70	Mumun	2430±50	BC 580

Table 5.9: The results of the AMS radiocarbon dating of the Songguk-Ri site

5.3.1. Sampling

5.3.1.1. Organic geochemical analysis

The samples for the organic geochemical analysis were collected during the 14th excavation of the Songguk-Ri site. The general sampling strategy for the site was somewhat different from that of the Kimpo-Yangchon and Sosa-Dong sites. Since the potsherds from the Songguk-Ri site were quite scarce, all the available ones which were conceded by the institution were sampled for the analysis. Under these circumstances, I collected a total of 27 samples from 16 house pits and 2 pit features (Table 5.10, Figure 5.20). If there are available radiocarbon dates from the house pits where the samples were collected, I have indicated in Table 5.10.

Sample No.	House pit or pit feature No.	Part	C^{14} date (BP; uncalibrated)
SON001	No. 52	Body	2560±40, 2460±40
SON002	No. 53	Body	
SON003	No. 54	Body	
SON004	No. 60	Body	
SON005	No. 70	Body	
SON006	No. 73	Body	
SON007	No. 77	Body	
SON008	No. 54 (pit feature)	Body	
SON009	No. 59 (pit feature)	Body	
SON010	No. 51	Body	2410±40, 2520±40
SON011	No. 51	Body	2410±40, 2520±40
SON012	No. 60	Body	
SON013	No. 60	Body	
SON014	No. 61	Body	
SON015	No. 72	Body	
SON016	No. 72	Body	
SON017	No. 74	Body	
SON018	No. 74	Body	
SON019	No. 52	Body	2560±40, 2460±40
SON020	No. 53	Body	
SON021	No. 58	Body	
SON022	No. 58	Body	
SON023	No. 59	Body	
SON024	No. 59	Body	

SON025	No. 62	Body
SON026	No. 63	Body
SON027	No. 69	Body

Table 5.10: The samples collected from the Songguk-Ri site for the organic geochemical analysis in this study

5.3.1.2. Luminescence dating

Unfortunately, no sample was collected for the Luminescence dating. This is due to the scarcity of potsherds unearthed during the 14th excavation.

5.3.2. Organic geochemincal results

Figure 5.21, 5.22, 5.23, and 5.24 show the results of the organic geochemical analyses. Among the 27 samples, 18 were analyzable. Nine samples were omitted due to contamination and the low concentration of lipids. Generally, the most frequently observed compounds in archaeological lipid residues are palmitic (C16:0) and stearic (C18:0) fatty acids (Evershed 2008a). The Songguk-Ri site was not an exception; and C16:0 and C18:0 fatty acids were the only organic compounds that were detected from all the analyzable 18 samples. Along with C16:0 and C18:0 fatty acids, I was able to identify both major short- and long-chain saturated fatty acids including C13:0, C14:0, C15:0, C17:0, C19:0, C20:0, C22:0, C23:0, C24:0. Again, most of the lipid residues analyzed showed a high concentration of C18:0 fatty acid (Figure 5.21). In addition to this pattern, some of the lipid samples displayed higher abundance of odd-numbered saturated fatty acids such as C15:0, C17:0, and C19:0 than others (Figure 5.21). According to the Evershed et al. (2002), these odd carbon number fatty acids are more abundant in ruminant animals than other monogastric organisms (Kwak et al., 2017: 7).

The geographical conditions of the Songguk-Ri site are not drastically different from the Kimpo-Yangchon and Sosa-Dong sites. Not too far away from the Songguk-Ri site is the Geum River, which is about 7 kilometers southwest of it. Therefore, aquatic resources might have had a chance to contribute to the diet of its dwellers. In order to fully understand whether these ancient farmers relied on aquatic resources, it is important to examine carefully the presence of aquatic biomarkers such as phytanic acid (3,7,11,15-tetramethylhexadecanoic acid), 4,8,12-TMTD (4,8,12-trimethyltridecanoic acid) and thermally produced long-chain ω-(o-alkylphenyl)alkanoic acids (cf. Craig et al., 2011; Evershed et al., 2008). Beside detecting phytanic acid from one sample (SON024), no other aquatic biomarkers were identified. Since phytanic acid is also found in the tissues of ruminant animals, we conclude phytanic acid alone from one sample is not a sufficient indicator of aquatic resource on its own (Heron and Craig, 2015). Despite the lack of fish bone remains and a rarity of aquatic biomarkers, we cannot exclude a possibility of fish as a part of diet in Songgukri culture, as it could have been cooked in a direct fire, leaving no mark on pottery (Kwak et al., 2017: 7).

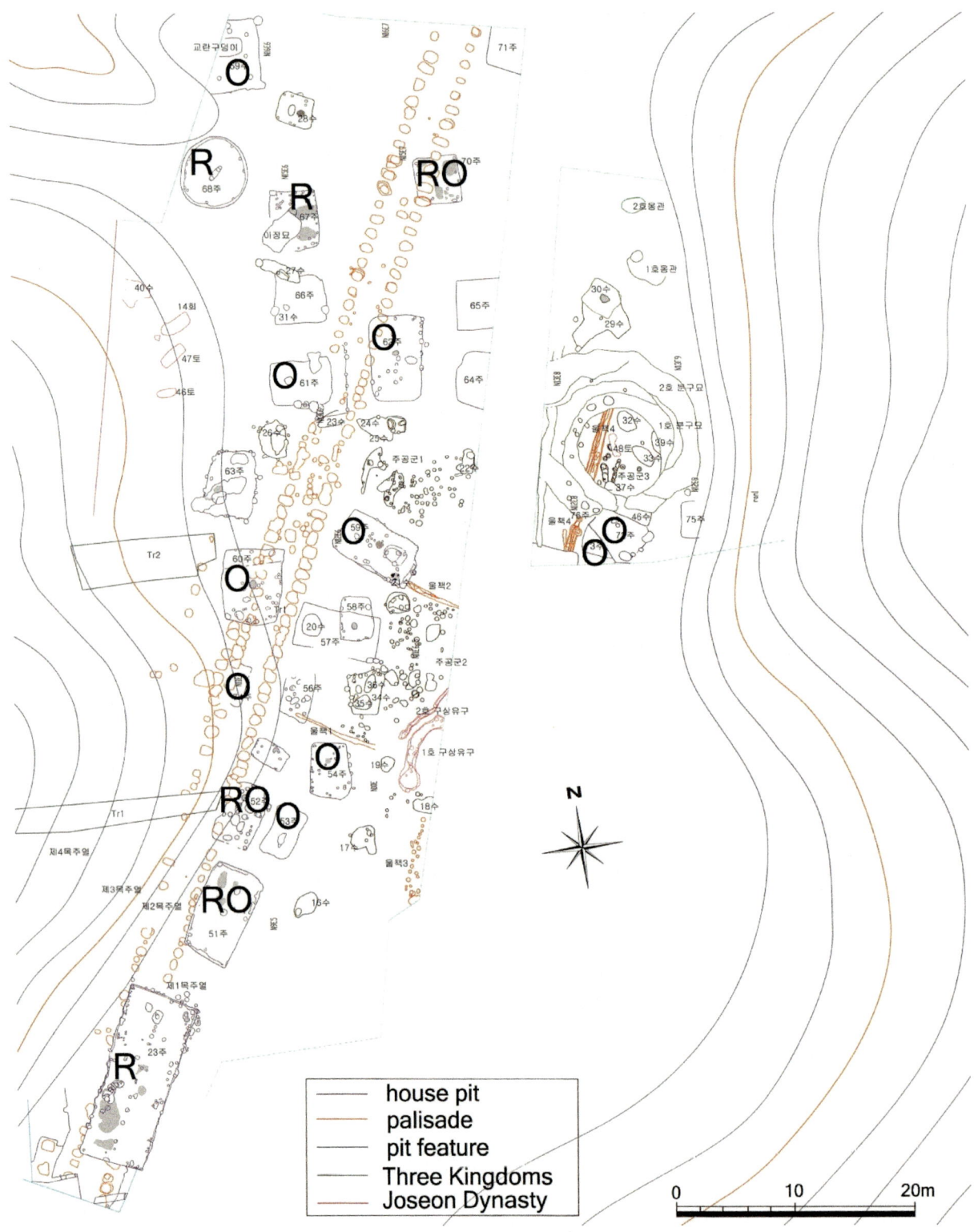

Figure 5.20: The site plan of the Songguk-Ri site and the location of the samples taken for the radiocarbon dating (R), organic geochemical analysis (O), and luminescence dating (L) (G. T. Kim et al., 2013)

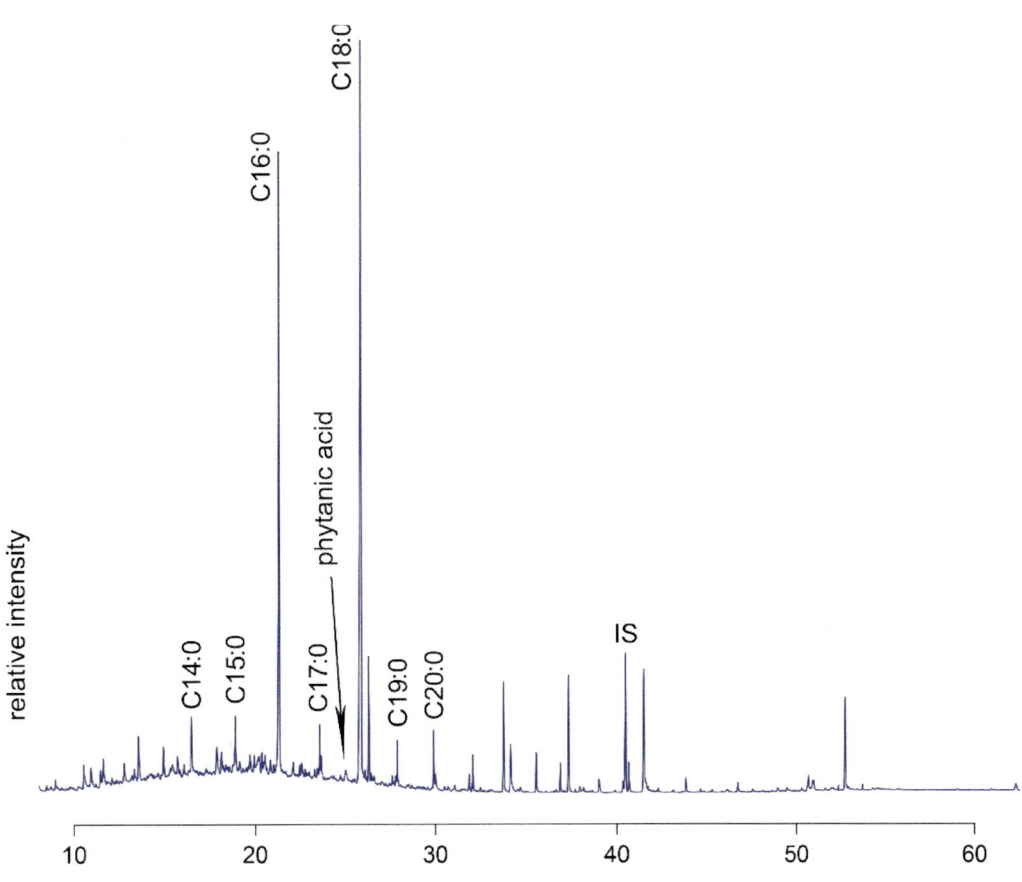

Figure 5.21: The result chromatogram of the GC-MS analysis of one of the samples from the Songguk-Ri site (SON024), using R version 3.2.0. Due to degradation, we usually observe medium- and long-chain saturated fatty acids. 5-α Cholestane was added as an internal standard (IS = 123.2 ng / microliter) (Kwak et al., 2017: 8)

The results of the isotope analyses of C16:0 and C18:0 fatty acids show their characteristic diet. At first glance, the story seems bit different from the former two cases. The results (Figure 5.22; 5.23; 5.24) indicate that they consumed several food stuffs including pork, C_3 plants and ruminants. Almost none of the samples indicated the presence of marine resources. This is probably because the distance between the site and the shore nearest to it is much farther than in case of the Kimpo-Yangchon and Sosa-Dong sites (Figure 5.2). Pork and ruminants were quite a popular foodstuff. Two samples indicated the presence of equine adipose. In Korea, the earliest confirmed evidence of domesticated horse came from several Late Mumun sites dated as early as 2300 BP (G. A. Lee, 2011; J. J. Lee, 2009). Considering that the Songguk-Ri site is classified into the Middle/Late Mumun period, it is quite possible that domesticated/wild horses would have contributed to its dwellers' diet.

During the 14th excavation of the Songguk-Ri site, over several thousands of carbonized grains were found. The dominant grains were foxtail millet and rice. Also, some of the samples analyzed in this study showed the presence of C_3 plant oils. However, since the area of C_3 plant oils in Figure 5.22 could indicate the mixture of pork and ruminant adipose (cf. Chapter four), we do not have any assurance that the C_3 plant oils of which the presence is indicated by those samples are actually plant oil. In addition to that, as I mentioned above, most of the lipid residues from Songguk-Ri showed a high concentration of C18:0 fatty acid, which generally indicates an animal source. However, this cannot exclude to possibility that harvested crops were processed of cooked in a pot at the Songguk-Ri site (Kwak et al., 2017: 8). Animal (and fish) fats are much more concentrated than plant oil in general and the former often mask the signal of the latter. Therefore, we are not removing the possibility that some (or most) of the pots were used for cooking both animals and grains.

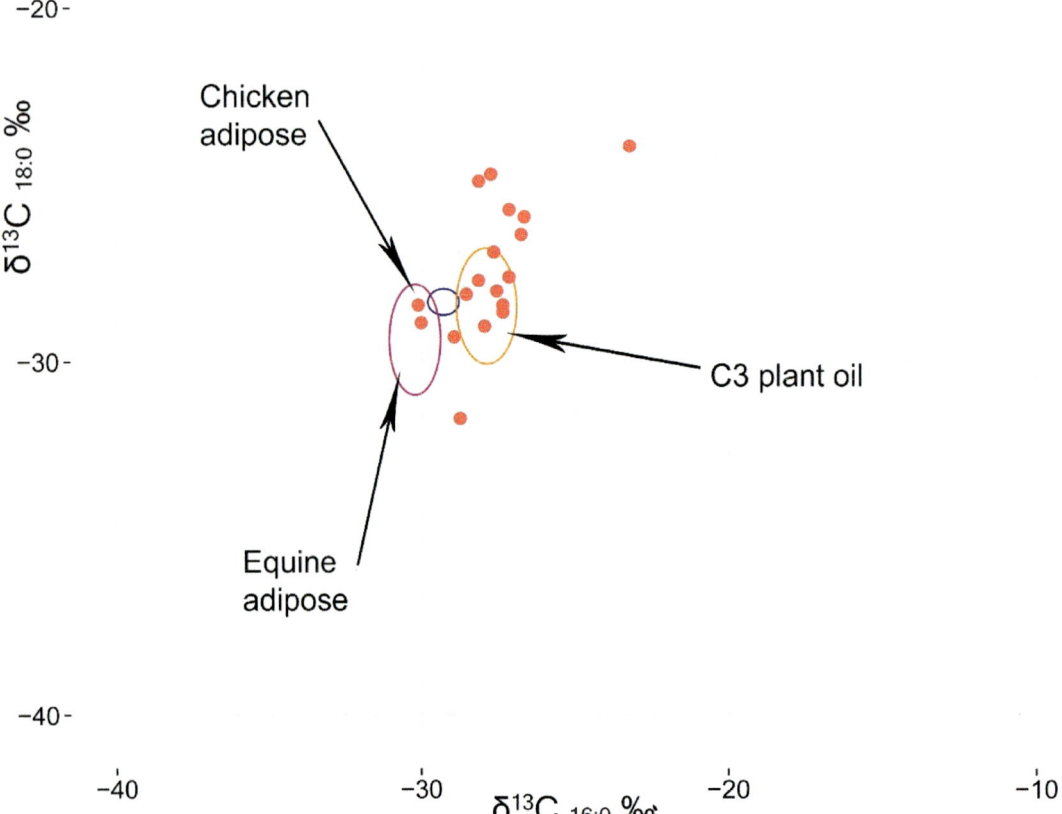

Figure 5.22: The results of CSIA by GC-C-IRMS of the samples from the Songguk-Ri site using the available references (cf. Dudd & Evershed, 1998; Dudd et al., 1999; Steele et al., 2010)

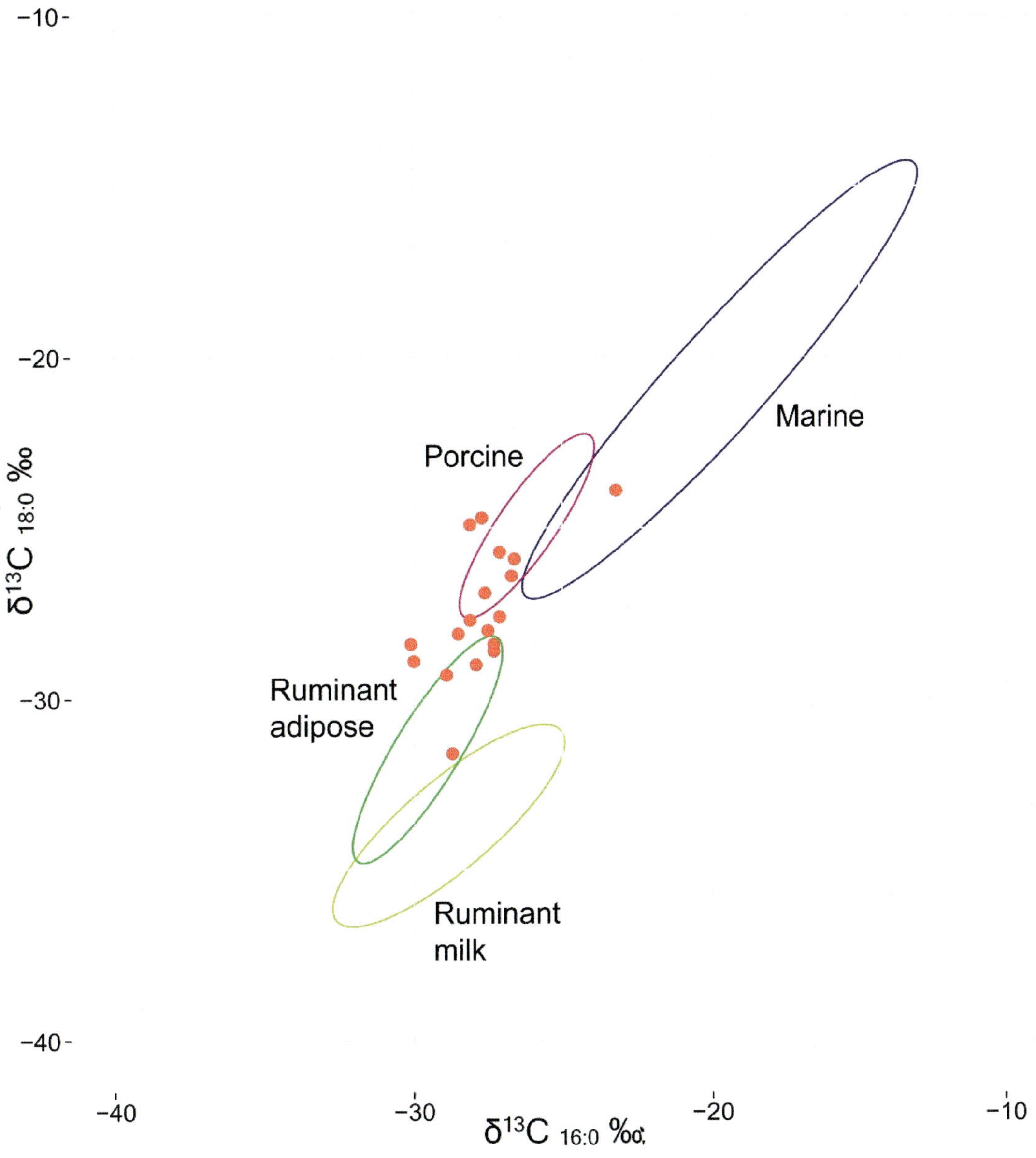

Figure 5.23: The results of CSIA by GC-C-IRMS of the samples from the Songguk-Ri site using the reference from Craig et al. (2011)

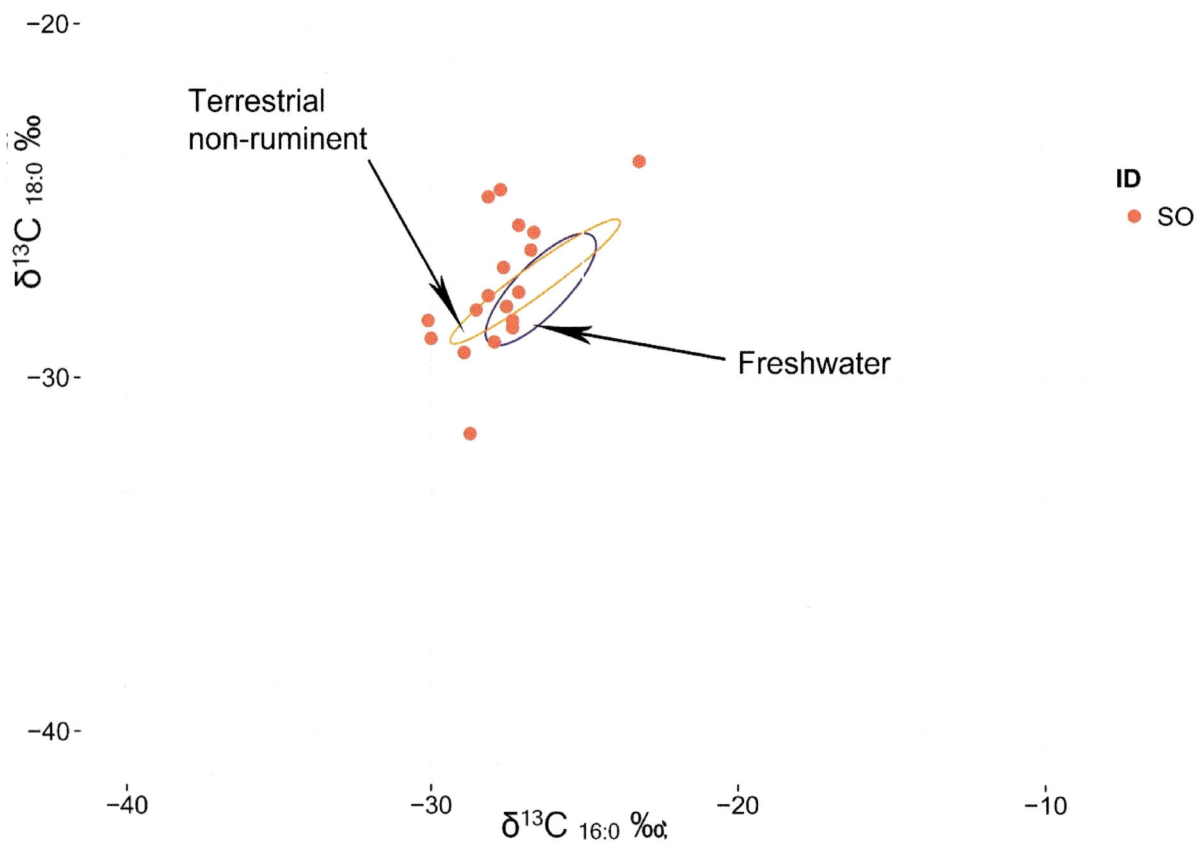

Figure 5.24: The results of CSIA by GC-C-IRMS of the samples from the Songguk-Ri site using the reference from Craig et al. (2013)

5.4. Eupha-Ri

Eupha-Ri is an Iron Age archaeological site in Huengseong city, Gangwon province, South Korea (Figure 5.2). The Huengseong city council had had a plan to build a cultural/athletic park; and the archaeological investigation had been performed beforehand by the Yonsei University Wonju Museum (H. J. Wang et al., 2013). The excavation was held from May 15th, 2009 to December 11th, 2011. The site contains various archaeological phenomena such as house pits, pit features and jar burials which represent different time periods from the Iron Age to the historical Joseon Dynasty (AD 1392 - 1897). The total site area is 23,840 square meters. Its main archaeological features belong to the Iron Age; and this thesis is focusing on this time period.

36 house pits, 24 pit features, and four jar burials were excavated and classified into the Iron Age. Based on AMS radio carbon dating applied to the four charcoal samples collected from the house pits, the main occupation period was assumed to be around 1,850 - 1,640 BP (H. J. Wang et al., 2013, Table 5.11, cf. Figure 5.26; 6.3). The house pits are either "呂" or "凸" shape with rounded corners, and contain interior features such as hearth and post holes (Figure 5.25). This description of their shape is based on the Chinese characters "Lu (呂)" and "Tu (凸)". The Iron age style hardened un-patterned pottery and that which was made by the beating method were excavated (Figure 5.27a). Other ceramic artifacts were also found, including a mold for iron casting, a net sinker and spindle whorls (Figure 5.27b). As for the Iron ware, axes, daggers and arrowheads were found (Figure 5.27b).

Overall, the Eupha-Ri site shows the typical characteristics of the Iron Age sites in the central part of the Korean Peninsula.

house pit No.	Cultural historical period	C^{14} date (BP; uncalibrated)	Calendar date
No. 1	Iron Age	1850±20	AD 188
No. 15	Iron Age	1780±20	AD 228
No. 15	Iron Age	1780±20	AD 217
No. 29	Iron Age	1640±20	AD 336

Table 5.11: The results of AMS radiocarbon dating of the Eupha-Ri site

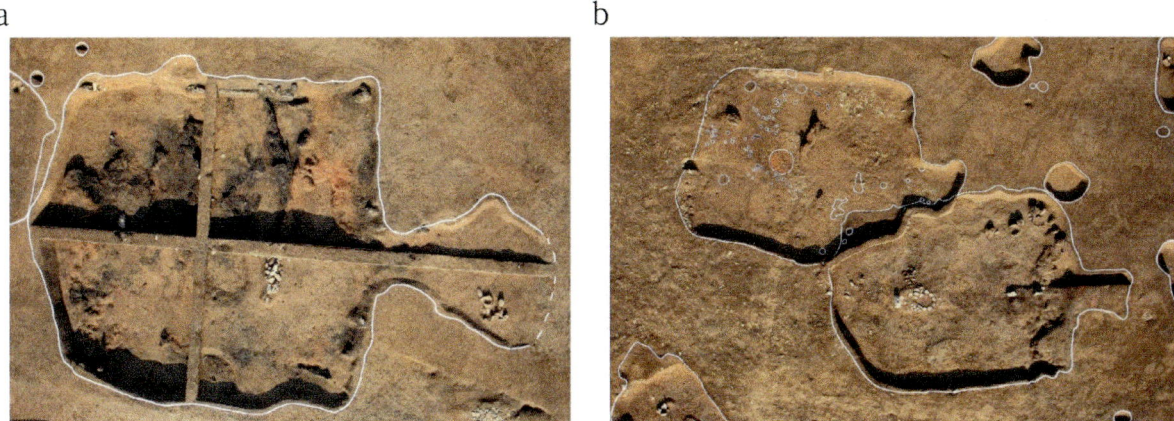

Figure 5.25: (a): Lu (呂) shape and (b): Tu (凸) shape house pits excavated from the Eupha-Ri site (Wang et al., 2013)

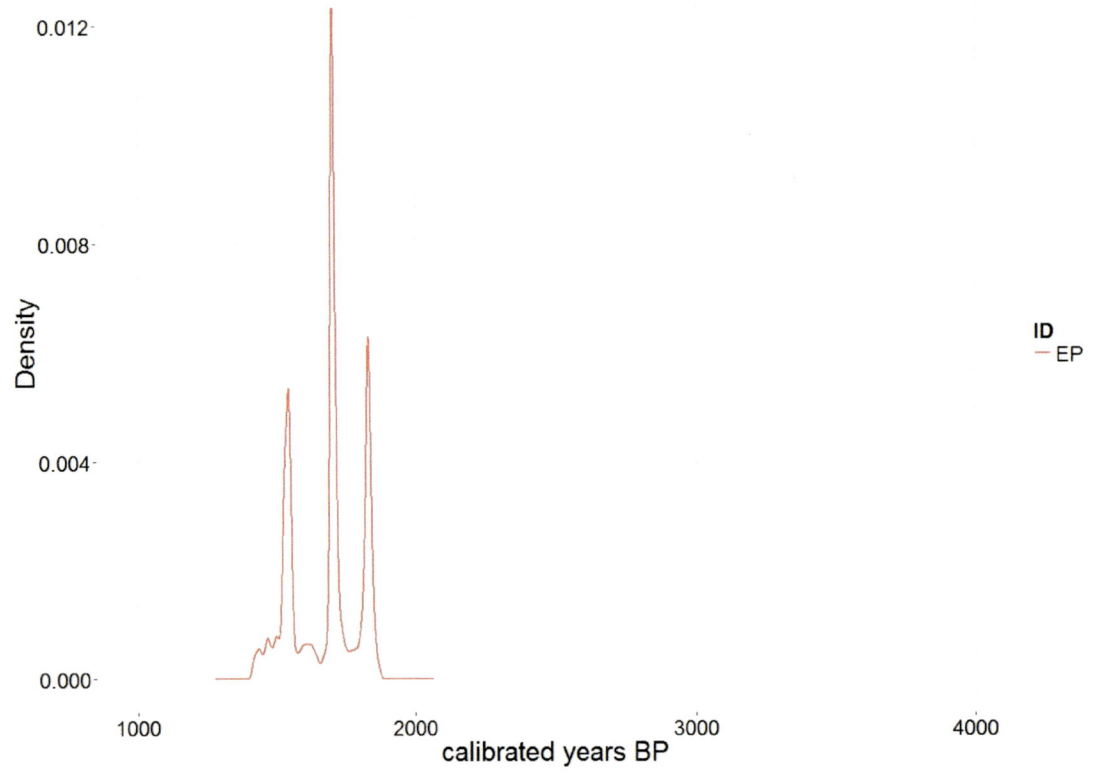

Figure 5.26: The density distribution of radiocarbon dates from the Eupha-Ri site, using the R package BChron (the dates were calibrated using the "intcal13" calibration curve, cf. Reimer et al., 2013)

Figure 5.27: Some of the artifacts uncovered during the excavation of the Eupha-Ri site (a): the Iron Age style hardened un-patterned pottery, a pot made by the beating method (center, second row) (b): mold for iron casting, net sinker, spindle whorls, iron axes and arrowheads (Wang et al., 2013)

5.4.1. Sampling

5.4.1.1. Organic geochemical analysis

Though numerous complete pots were excavated (Figure 5.27a), not many potsherds were found. Since in the archaeological investigation, the priority is given to preserving pots in their original form and since it is not common to find lots of complete ones, I was not allowed to take parts from complete ones for the analyses. Under these limited conditions, the samples were collected among the available potsherds found at house pits. Thus, a total of 25 samples were collected from eight house pits (Table 5.12, Figure 5.28). Though I tried to collect as many samples as I could in the given situation, I have to confess that the eight house pits might not fully represent the entire aspect of the site. If there are available radiocarbon dates from the house pits where the samples were collected, I have indicated them in Table 5.12.

Sample No.	House pit No.	Part	C^{14} date (BP; uncalibrated)
EUP001	No. 15	Rim	1780±20
EUP002	No. 15	Rim	1780±20
EUP003	No. 15	Rim	1780±20
EUP004	No. 15	Rim	1780±20
EUP005	No. 15	Rim	1780±20
EUP006	No. 15	Rim	1780±20
EUP007	No. 15	Rim	1780±20
EUP008	No. 15	Bottom	1780±20
EUP009	No. 15	Bottom	1780±20
EUP010	No. 15	Bottom	1780±20
EUP011	No. 15	Bottom	1780±20
EUP012	No. 33	Rim	
EUP013	No. 32	Body	
EUP014	No. 32	Body	
EUP015	No. 32	Body	
EUP016	No. 29	Body (beating method)	1640±20
EUP017	No. 15	Body	1780±20
EUP018	No. 15	Rim	1780±20
EUP019	No. 7.8.9 disturbed	Rim	
EUP020	No. 7.8.9 disturbed	Rim	
EUP021	No. 12	Rim (beating method)	
EUP022	No. 7.8.9 disturbed	Rim (beating method)	
EUP030	No. 33	Body	
EUP031	No. 32	Body	
EUP032	No. 29	Body	1640±20

Table 5.12: The samples collected from the Eupha-Ri site for the organic geochemical analysis in this study

5.4.1.2. Luminescence dating

For the luminescence dating three samples were collected. Among the three samples, one was collected from the house pit that had been dated by the radiocarbon dating, and the other two from those which had not been dated (Table 5.13, Figure 5.28).

Sample No.	Location/house pit No.	Part	Depth (m)
U3039	No. 33	Body	0.3
U3040	No. 32	Body	0.3
U3041	No. 29	Body	0.3

Table 5.13: The samples collected from the Eupha-Ri site for the luminescence dating in this study

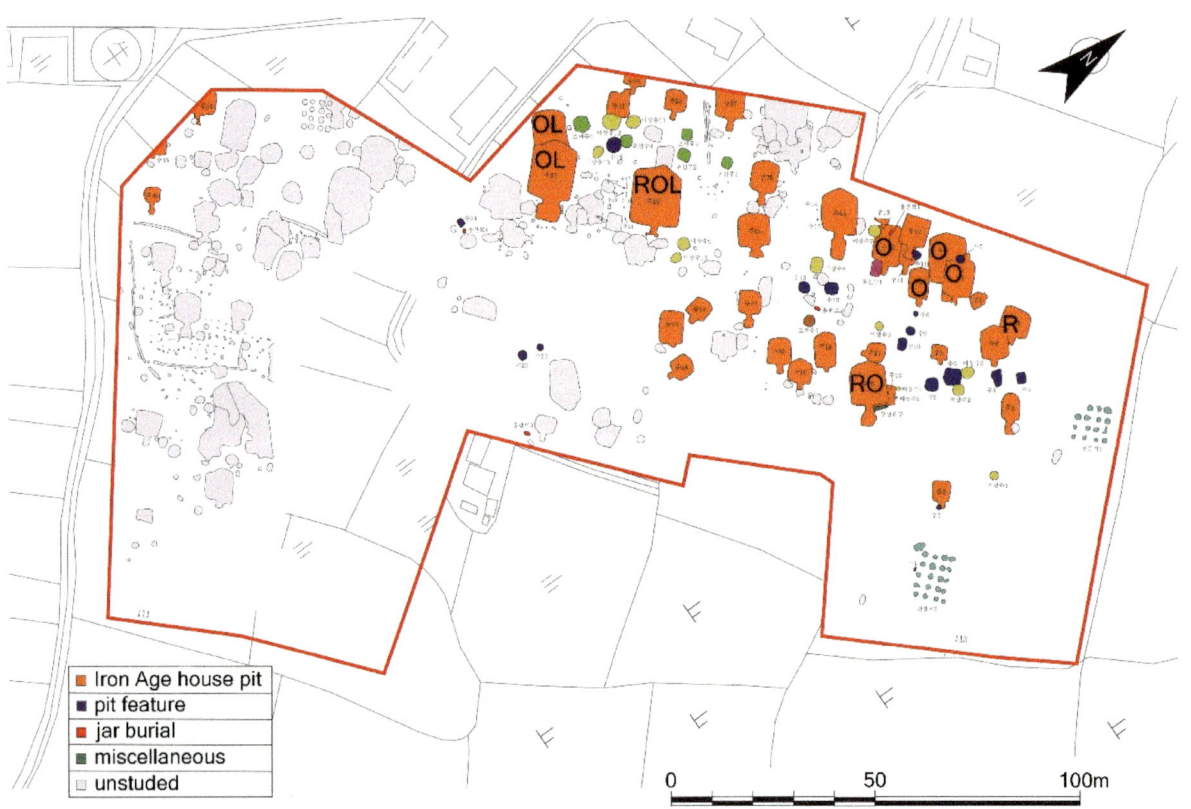

Figure 5.28: The site plan of the Eupha-Ri site and the location of the samples taken for the radiocarbon dating (R), organic geochemical analysis (O), and luminescence dating (L) (H. J. Wang et al., 2013)

5.4.2. Organic geochemical results

Figure 5.29, 5.30, 5.31 and, 5.32 show the results of the organic geochemical analyses. Among the 25 samples, only eight were analyzable. 17 samples were omitted mostly due to the low concentration of lipids. Like the results in case of the former three sites, palmitic (C16:0) and stearic (C18:0) fatty acids were detected from all the analyzed eight samples. Along with C16:0 and C18:0 fatty acids, I was able to identify both major short- and long-chain saturated fatty acids such as C14:0, C15:0, C17:0, C19:0, C20:0, C21:0, C22:0, C22:0, C23:0 and C24:0. The overall lipid concentration of the samples from the Eupha-Ri site was quite low; and the number of the identified fatty acids was much smaller than those at the former three sites.

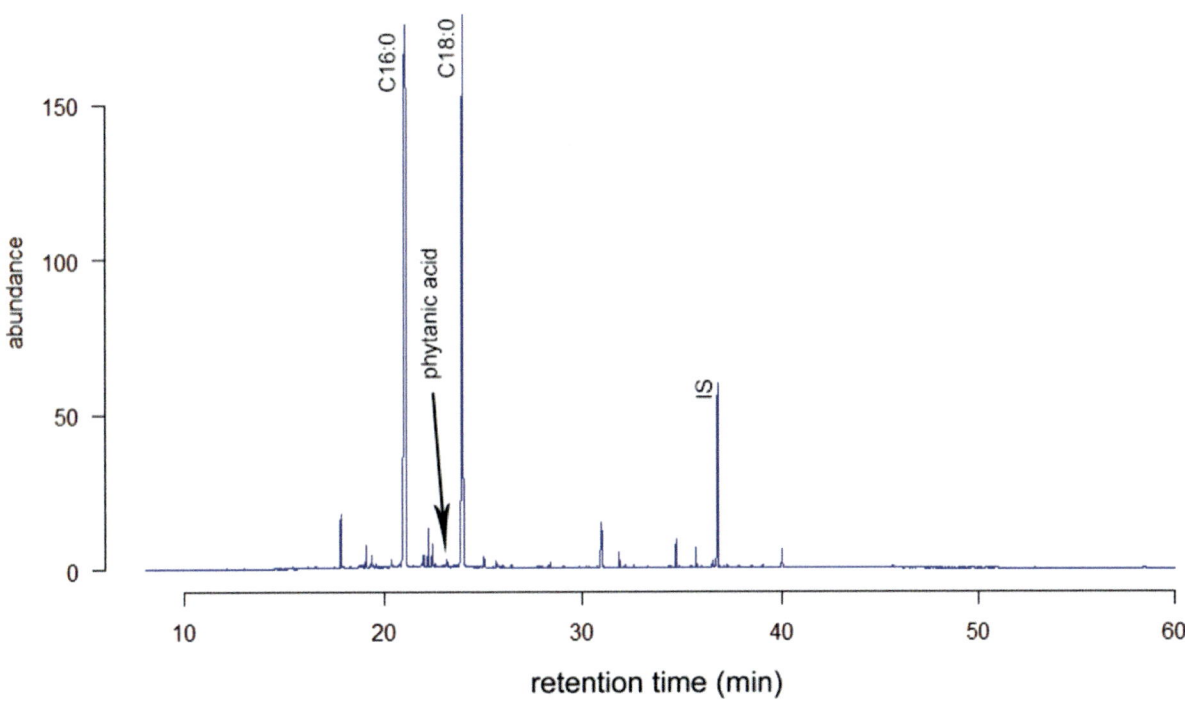

Figure 5.29: The result chromatogram of the GC-MS analysis of one of the samples from the Eupha-Ri site (EUP005), using R version 3.2.0. Due to degradation, we usually observe medium- and long-chain saturated fatty acids. 5-α Cholestane was added as an internal standard (IS = 44 ng / microliter)

This was quite striking, because the Eupha-Ri site is almost 1000 years younger than the other sites (such as Kimpo-Yangchon or Sosa-Dong), and I thought lipids in younger sites had more chances to survive against the post-depositional processes than in older ones. The overall low concentration of lipids at the Eupha-Ri site is probably due to the hard fabric of the Iron Age potteries. The surface treatments and a high firing temperature brought into play in manufacturing the Iron Age ceramic vessels would have generated smaller pores, which would have limited the concentration of lipids (cf. Correa-Ascencio & Evershed, 2014). Otherwise, more lipids could have been absorbed into the vessels. Though Correa-Ascencio and Evershed (2014) showed the effectiveness of the methanolic acid extraction on hard and burnished pots, the lipid concentration of the Eupha-Ri site's potsherds was still low, compared with that which had been observed at more porous Mumun potteries. Most of the lipid residues analyzed showed a high abundance of C18:0 fatty acid (Figure 5.29).

Geographically, the Eupha-Ri site is just near the Seom River. Therefore, aquatic resources, especially fresh water ones, might have had a chance of having contributed to its dwellers' diet. In order to fully understand whether they relied heavily on aquatic resources or not, it is important to carefully examine the presence of aquatic biomarkers such as phytanic acid (3,7,11,15-tetramethylhexadecanoic acid), 4,8,12-TMTD (4,8,12-trimethyltridecanoic acid), and thermally produced long-chain ω-(o alkylphenyl)alkanoic acids (cf. Craig et al., 2011; Evershed et al., 2008). Among the eight samples, one sample showed the presence of phytanic acid (EUP005).

In the Eupha-Ri site, the diet pattern is somewhat different from that of the former three cases. The isotope analyses of C16:0 and C18:0 fatty acids shows its interesting aspect. The results of the analyses (Figure 5.30; 5.31; 5.32) indicate that the site's ancient dwellers mainly consumed Ruminants. This diet pattern focused on ruminants in the Iron Age is quite different from that of the Mumun period in which pork was also major part of the diet.

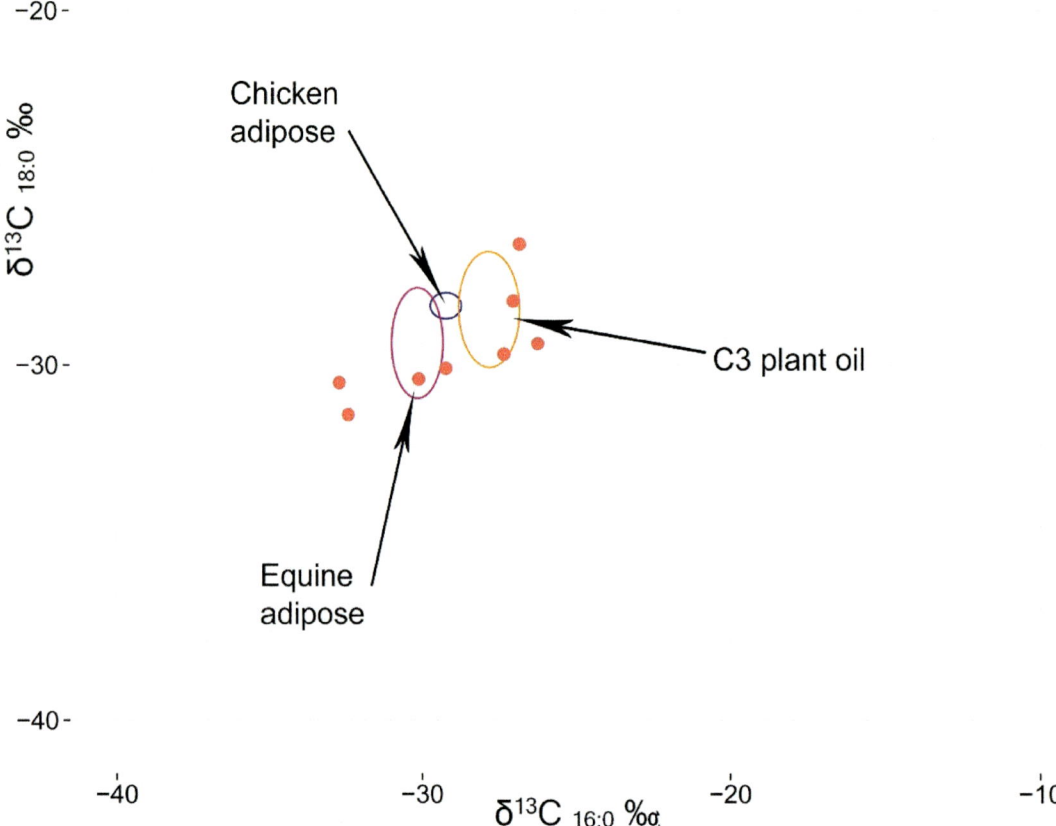

Figure 5.30: The results of CSIA by GC-C-IRMS of the samples from the Eupha-Ri site using the available references (cf. Dudd & Evershed, 1998; Dudd et al., 1999; Steele et al., 2010)

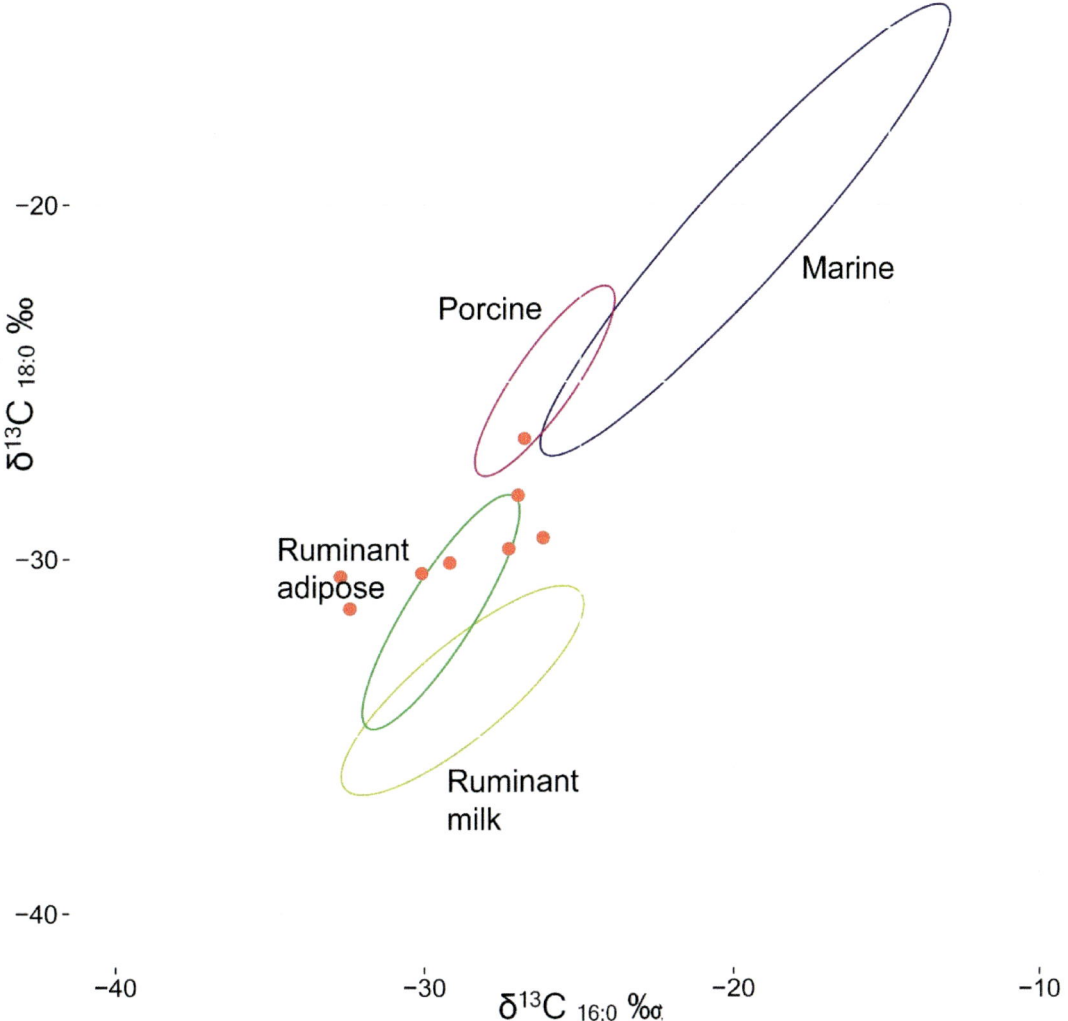

Figure 5.31: The results of CSIA by GC-C-IRMS of the samples from the Eupha-Ri site using the reference from Craig et al. (2011)

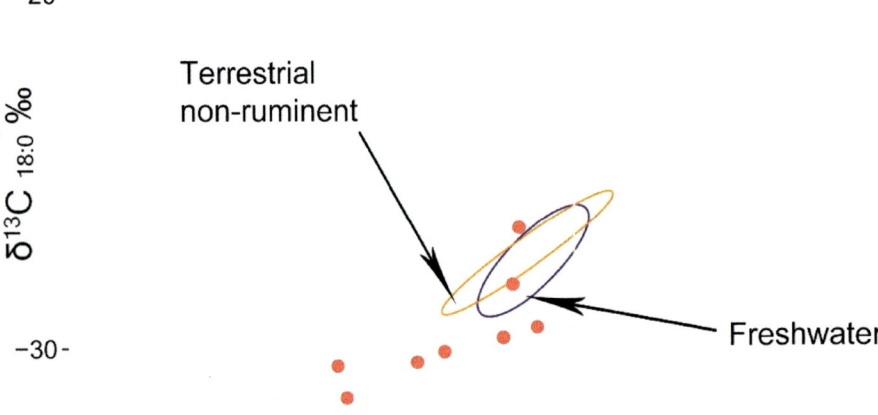

Figure 5.32: The results of CSIA by GC-C-IRMS of the samples from the Eupha-Ri site using the reference from Craig et al. (2013)

5.4.3. Luminescence dating results

The samples were dated using TL, OSL and IRSL at the luminescence dating lab, University of Washington. Table 5.14 shows the results of the luminescence dating. The OSL signal for UW3039 was most likely from quartz. The IRSL signal was weak and the OSL b-value was obviously in the range of quartz. The TL age was generally older but only by about 120 years. The OSL age gave the best estimate for the sample UW3039. The ages for OSL, IRSL, and TL were all in agreement for the sample UW3040 (the fading was not significant). The ages for OSL and IRSL were in agreement for the sample UW3041. The TL age was younger probably due to the very high fading rate. U3039 and U3041 corresponded to the published four AMS radiocarbon dates (Table 5.11). The date presumed by U3041 indicates that the site was occupied by the Iron Age people slightly longer than the radiocarbon dates suggest. The result of U3040 did not match with both the archaeological features of the site and the radiocarbon dates.

Lab. No	Depth (m)	Water Content (%)	Dose rate* (Gy/ka)	De (Gy)			Age
				TL	OSL	IRSL	
U3039	0.30	14	6.378±0.429	14.67±1.07	8.13±0.264	14.652±2.537	160±120 AD
U3040	0.30	13.4	5.336±0.562	12.3±3.1	8.974±0.16	10.04±0.214	260±110 BC
U3041	0.30	12.6	6.596±0.300	7.916±0.598	8.578±0.186	8.534±0.326	530±60 AD

Table 5.14: The results of the luminescence dating of the potsherd samples from the Eupha-Ri site. The overall low water content of the samples shows the less porous nature of the Iron Age pottery.

5.5. Isotope implication: Alternative approach

As I mentioned in chapter four, in this study, the stable carbon isotope values of C16:0 and C18:0 fatty acids from the archaeological samples were compared with the available modern references that were obtained from the modern fauna and flora that exist in either Japan, Northern Europe or North America (Copley et al., 2003; Craig et al., 2013, 2011; Cramp et al., 2011; Dudd et al., 1998, 1999; Evershed et al., 1994, 1997; Mottram et al., 1999; Reber & Evershed, 2004a; Steele et al., 2010) to detect the presence of the potentially cooked resources in the prehistoric Korean Peninsula. Although the overall ecosystem of Japan, Northern Europe, and North America is similar to that of Korea, this approach cannot be separated from the inherent assumption that the $\delta^{13}C$ values of available modern samples collected from those regions are comparable to archaeological ones from the Korean Peninsula.

Salque et al. (2013)'s approach can remove the exogenous factors related to the local environment (e.g. C_4 plants contribution) and allows us focus on the metabolic and biosynthetic characteristics of the animal fat source (Copley et al., 2003; Dunne et al., 2012). This approach uses the $\Delta^{13}C$ ($\delta^{13}C_{18:0} - \delta^{13}C_{16:0}$) proxy which can separate fats based on physiological differences between the animals regardless of differences of their diets or surrounding ecosystem (cf. Copley et al., 2003; Evershed, 2008b, cf. Figure 4.6). With this approach, we can make distinction between non-ruminant adipose, ruminant adipose, and ruminant milk. Most of the lipid residues extracted from the analyzable samples in this study showed high concentration of C18:0 fatty acid (cf. Figure 5.5; 5.14; 5.21; 5. 29). Among them, quite a few samples displayed higher abundance of odd-numbered saturated fatty acids including C15:0, C17:0, and C19:0 (Figure 5.21). All of these generally indicate animal sources including ruminants (Enser 1991; Copley et al. 2005c; Evershed et al. 2002: 664; Kwak at al., 2017: 7). In this setting, I decided to employ the method of Salque et al. (2013) for the interpretation of CSIA.

Figure 5.33, 5.34, and 5.35 shows the CSIA results of the all four sites in this study using the interpretational approach of Salque et. al. (2013). Almost all of the samples indicated the presence of the ruminant and non-ruminant adipose, but none of the samples were used for processing milk fats (one sample from the Sosa-Dong in Figure 5.33b is mechanical outlier). This general trend is more or less similar to the patterns that we were able to witness during the interpretation of CSIA based on the modern references from Northern Europe, Japan, or North America throughout this chapter. Above all, one clear finding is that terrestrial animals (including ruminants) were consumed on a regular basis at the all four sites examined in this study.

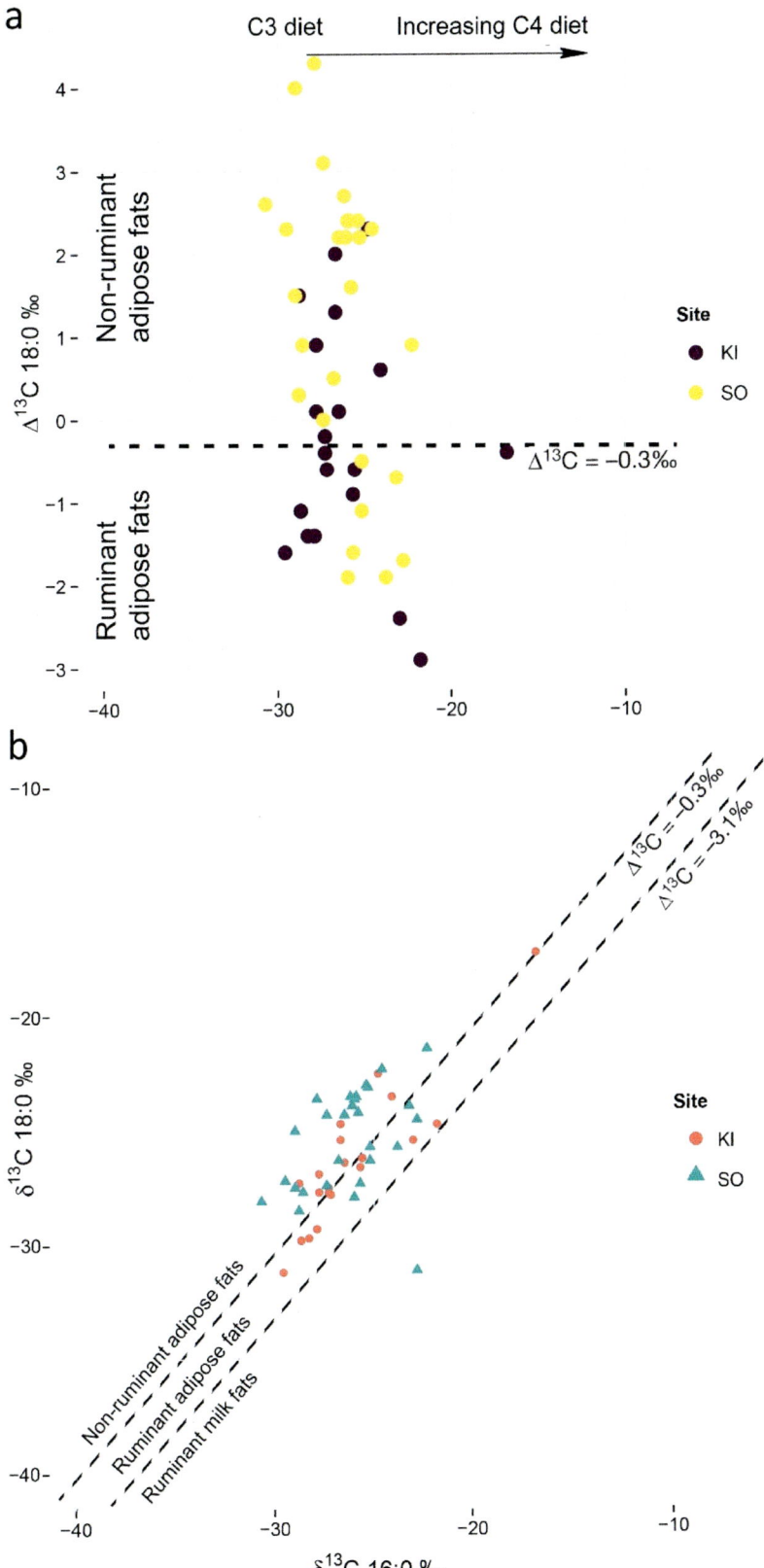

Figure 5.33: (a, b) The results CSIA of the samples from the Kimpo-Yangchon and Sosa-Dong site using the approach of Salque et al. (2013). (a) The $\delta^{13}C$ values suggest a C_3 biased across the samples, indicating terrestrial herbivores mainly consumed indigenous wild C_3 plants (cf. Ahn, 2006; Choy and Richards, 2010; J. J. Lee, 2011b).
$\Delta^{13}C$ 18:0 = $\delta^{13}C_{18:0}$ - $\delta^{13}C_{16:0}$, KI = Kimpo-Yangchon, SO = Sosa-Dong

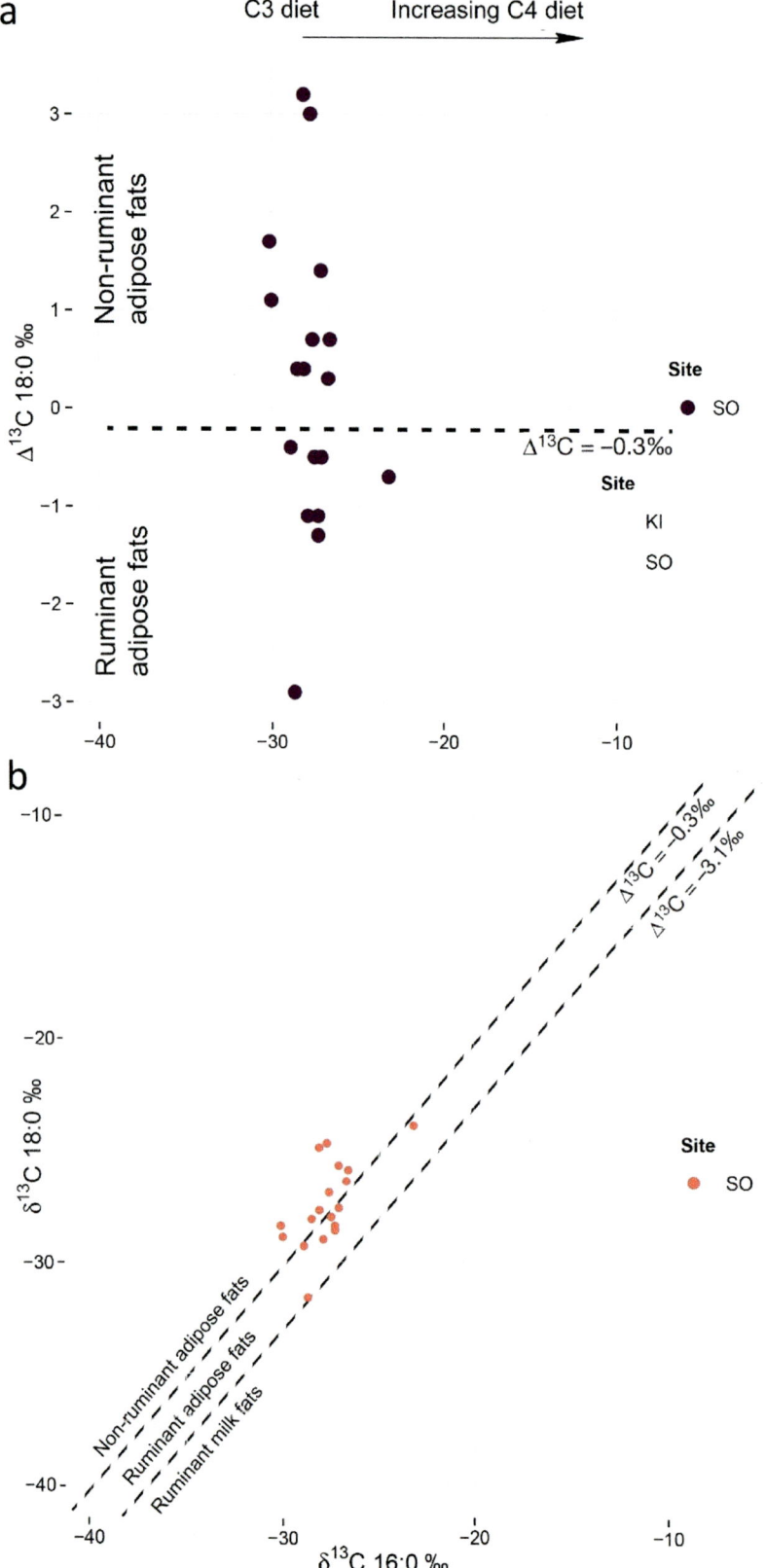

Figure 5.34: (a, b) The results CSIA of the samples from the Songguk-Ri site using the approach of Salque et al. (2013) (Kwak et al., 2017: 9). (a) The $\delta^{13}C$ values suggest a C_3 biased across the samples, indicating terrestrial herbivores mainly consumed indigenous wild C_3 plants (cf. Ahn, 2006; Choy and Richards, 2010; J. J. Lee, 2011b). $\Delta^{13}C$ 18:0 = $\delta^{13}C_{18:0}$ - $\delta^{13}C_{16:0}$, SO = Songguk-Ri

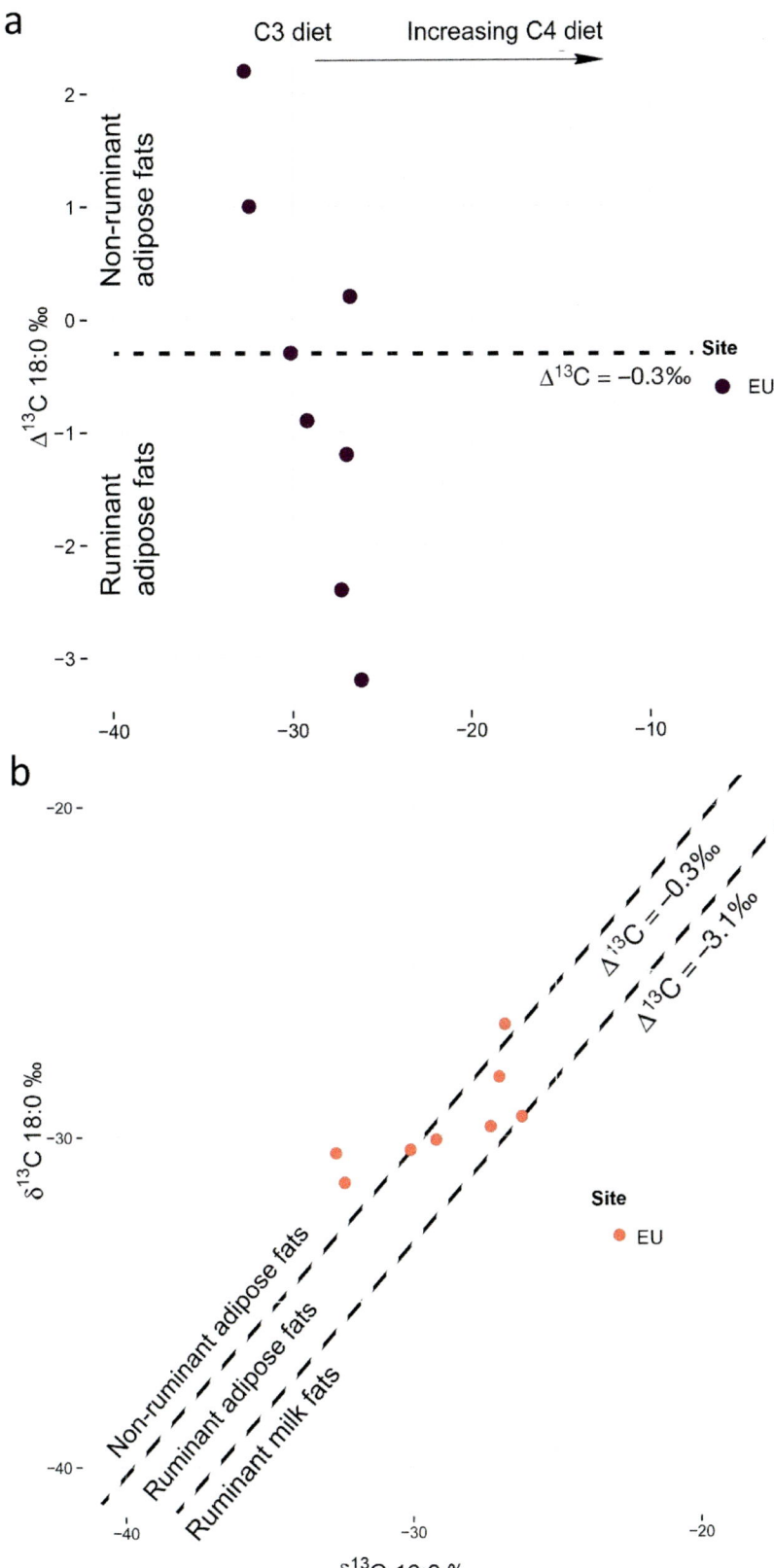

Figure 5.35: (a, b) The results CSIA of the samples from the Eupha-Ri site using the approach of Salque et al. (2013). (a) The $\delta^{13}C$ values suggest a C_3 biased across the samples, indicating terrestrial herbivores mainly consumed indigenous wild C_3 plants (cf. Ahn, 2006; Choy and Richards, 2010; J. J. Lee, 2011a; 2011b). $\Delta^{13}C$ 18:0 = $\delta^{13}C_{18:0} - \delta^{13}C_{16:0}$, EU = Eupha-Ri

6. Discussion

6.1. The subsistence of the Chulmun and Mumun periods

In chapter two, we have observed the connection between Chulmun and Mumun through the examination of changes in pattern on the potteries (cf. Figure 2.8). What, then, about the subsistence of the Chulmun and Mumun peoples? Were the foodstuffs of the Chulmun people quite different from those of the Mumun people? What was the role of rice? Did the Mumun people really rely heavily on rice and harvested crops as previously suggested? Was rice a hallmark of the Mumun period?

6.1.1. The Chulmun subsistence

Traditionally, what we know about the subsistence of the Chulmun people is that they relied mainly on hunting and gathering; but from the Middle stage (5,500 BP) of the Chulmun period, we are able to observe evidence of the initial domestication of foxtail and broom-corn millet (Norton, 2007). Recently. G. Lee (2011: p. S326) argued that they had specific subsistence solutions which include combinations of wild (e.g. acorn (*Quercusacutissima*Carr.), Manchurian walnut (*Juglans* spp.)), possibly managed (e.g. chenopod (*Chenopodium* sp.), panicoid grass (*Paniceae*)), and domesticated (e.g. foxtail (*Setariaitalica* ssp. *italica*) and broomcorn millet (*Panicummiliaceum*), possibly soybean (*Glycine max*), azuki (*Vignaaugularis*) and beefsteak plant (*Perillafrutescens* (L.) Britt)) plants. However, before we define the subsistence of the Chulmun period, it is worth to examine available bulk isotope data.

Throughout the Chulmun period, ancient people occupied the coastline of almost the entire Korean Peninsula, and created shell middens in many different locations near coastal areas. Because of these shell middens, it is easier for us to trace the ancient diet, for they provide excellent environment in terms of preservation of organic materials. Numbers of bulk Isotope studies have been conducted, since both human and animal bones were excavated from these middens (D. I. Ahn, 2006; Choy & Richards, 2010; H. S. Kim, 2010).

The isotope analysis on the human remains and animal bones excavated from the Geoje and Tongsam-Dong shell middens, southern part of the Korean Peninsula (Figure 2.7) shows a focused diet of the Chulmun people and terrestrial animals (Choy & Richards, 2010; H. S. Kim, 2010). The relatively low $\delta^{13}C$ and $\delta^{15}N$ values of wild terrestrial mammals such as pigs and deer indicate that the diets of these animals were dominated by C_3 plants (Figure 6.1), which means most of indigenous wild plants in the Korean Peninsula are C_3 plants (cf. J. J. Lee, 2011b). On the other hand, the $\delta^{13}C$ and $\delta^{15}N$ values of the human bone collagen were quite close to those of marine animals (Figure 6.1), indicating the Chulmun people mainly consumed marine resources. It suggests, along with the geographic location of the middens (Figure 2.7), that procuring marine resources was their main subsistence strategy, though terrestrial mammals were included in their diet.

D. I. Ahn (2006, Figure 6.2) conducted isotope analysis on the human remains and animal bones for the Konam-Ri shell midden (Figure 2.1b), central part of the Korean Peninsula. The samples were collected from both the Chulmun and Mumun periods. The isotope analysis on pig bones from both periods indicates that the pigs mainly consumed C_3 plants. The $\delta^{13}C$ and $\delta^{15}N$ values of the human bone collagen from the Chulmun period were close to those of a wild pig, which means the Chulmun people mainly relied on wild C_3 plants as well as wild pigs. However, considering the difference in $\delta^{13}C$ values between human (-17.7 ‰) and pig (-19.7 to -21.2 ‰), we cannot totally eliminate the possibility of C_4 plants in the human diet.

Based on the results of isotope analyses and paleobotanical studies, one may conclude that the Chulmun people mainly relied on hunted terrestrial animals/marine mammals, gathered wild C_3 plants and, in some degree, domesticated C_4 plants.

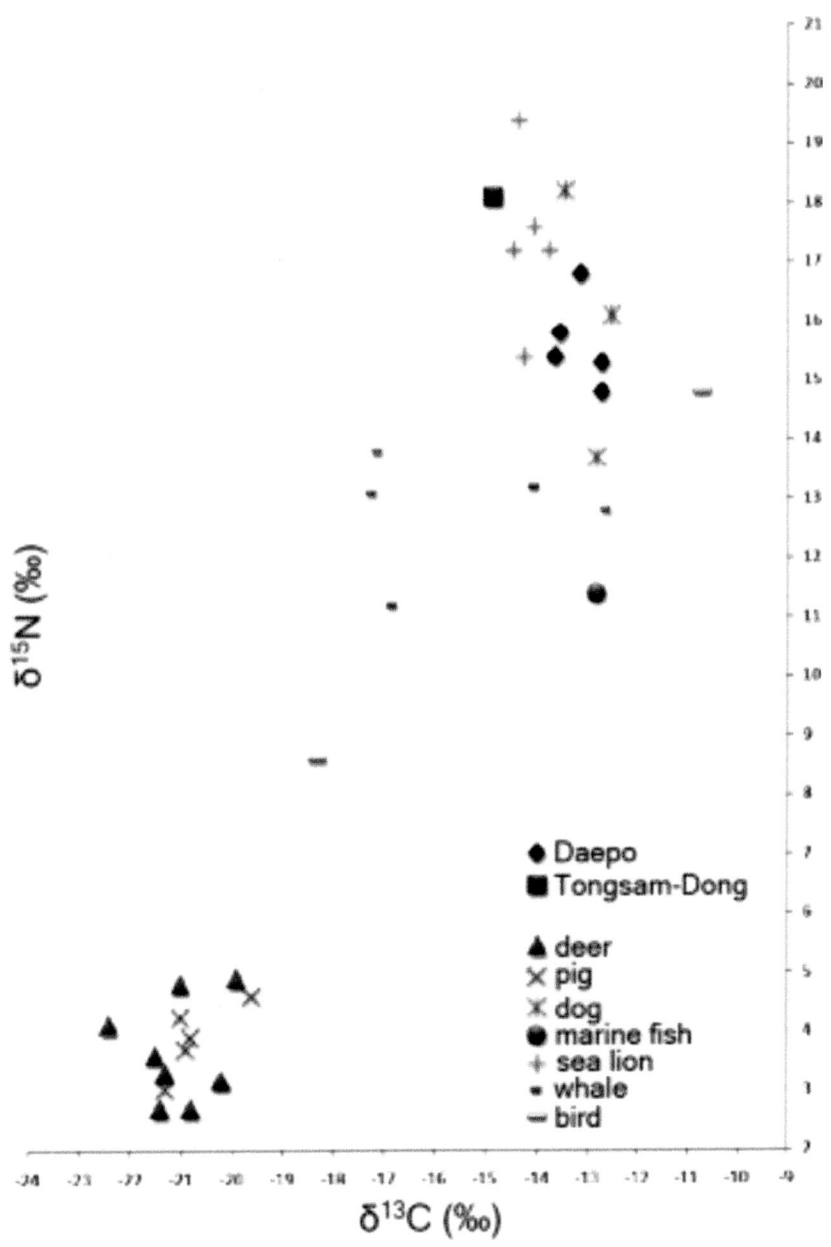

Figure 6.1: The results of the bulk isotope analysis on human remains and animal bones excavated from the Daepo and Tongsam-Dong shell middens (modified from J.J. Lee 2011b: p. 41)

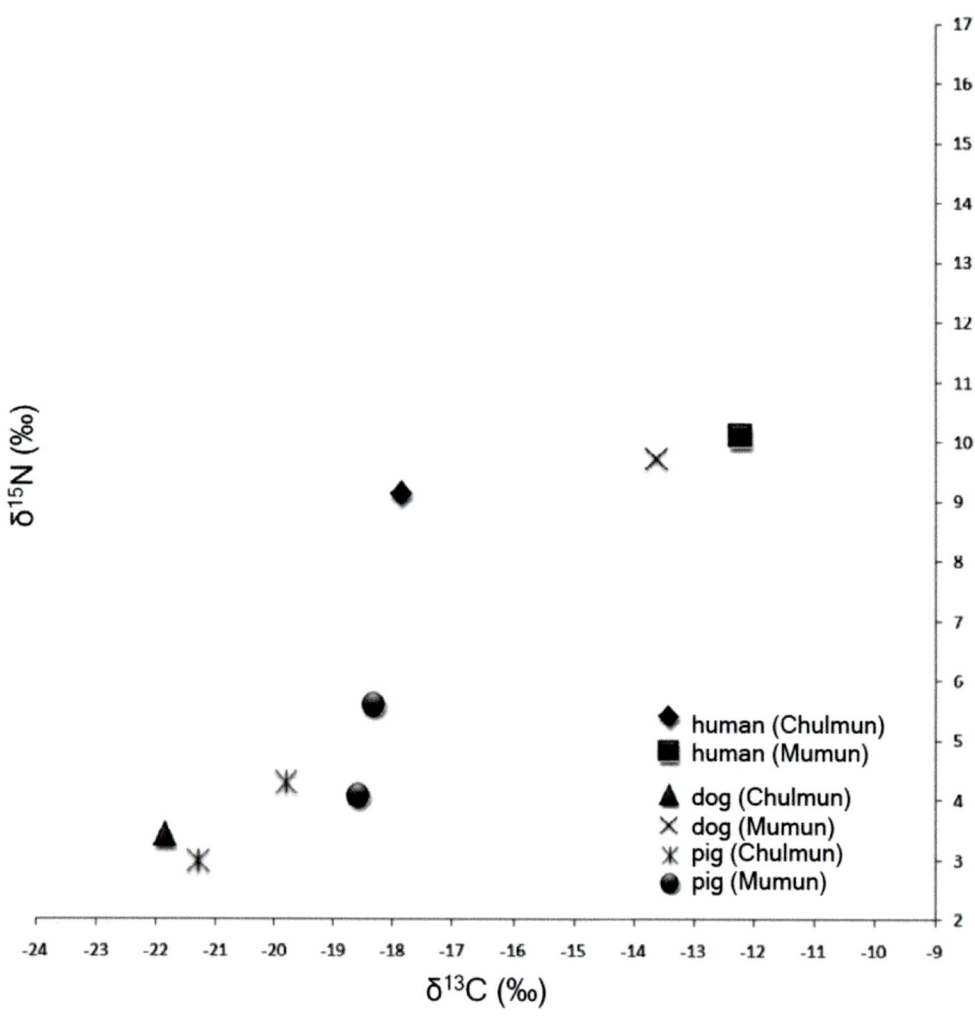

Figure 6.2: The results of the bulk isotope analysis on human remains and animal bones excavated from the Konam-Ri shell middens (modified from J. J. Lee, 2011b: p. 44)

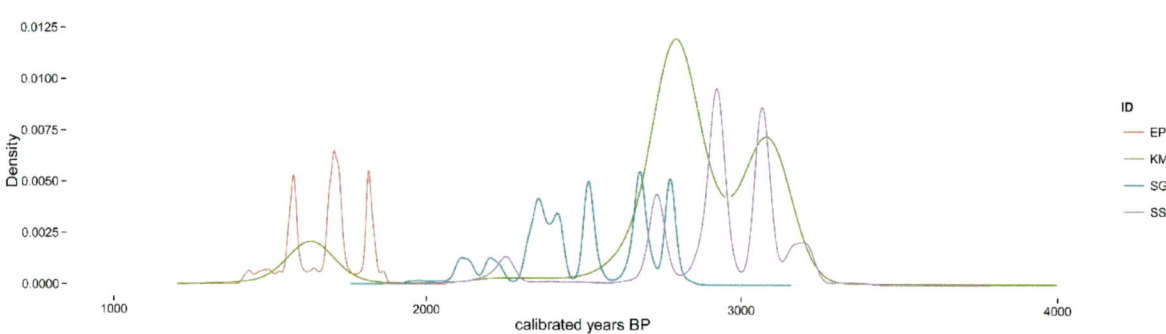

Figure 6.3: Density distributions of all radiocarbon dates from each site studied in this thesis, using the R package BChron (Sosa-Dong: SS, Kimpo-Yangchon: KM, Songguk-Ri: SG, Eupha-Ri: EP) All dates were calibrated using 'intcal13' calibration curve

6.1.2. The Mumun subsistence

One of the most debated issues related to the Mumun subsistence in the central part of the Korean Peninsula is the role of the intensive agriculture heavily based on rice. On one hand, the intensive agriculture was viewed as "cure-all remedy" (G. A. Lee, 2011: p. S327) which substituted for any other subsistence resources (J. S. Kim, 2003, 2006; J. J. Lee, 2001; Norton, 2000). In addition to that, the emergence of a social hierarchy and the subsequent social complexity were considered to be driven by the rapid spread of the intensive rice agriculture into foraging contexts (B. C. Kim, 2006a, 2006b). On the other hand, the role of both broom corn and foxtail millet was emphasized together with that of azuki and soybean (Crawford & Lee, 2003; G. A. Lee, 2011). In the latter view represented by paleobotanical studies, contrary to the former, rice agriculture was regarded as no more than an 'add on' to the existing millet-based subsistence originating from the Chulmun period. However, the results described in the previous chapter of the organic geochemical analyses of potsherds from major Mumun habitation sites told a story somewhat different from either of these existing arguments.

At least what we have learned from the organic geochemical analyses throughout chapter five, terrestrial animals including pork and ruminants occupied major part of the Mumun farmers' diet (3000 - 2300 BP, Figure 6.3). Most of the lipids analyzed showed the characteristics of the animal origin (cf. chapter five). However, this is not a surprising result, for we already have solid evidence that sika deer (Cervus Nippon) and wild boar (Sus scrofa) were the main protein source since the Neolithic period (Choy et al., 2012; J. J. Lee, 2011a; 2011b; Shin et al., 2013; cf. Kwak et al., 2017). In a recent study, J. J. Lee (2011a) even mentioned the symbolic significance of pork consumption in the Korean Peninsula. Whether or not there are symbolic meanings in pork consumption in these samples, it is unquestionable that boar hunting was one of the major subsistence strategies since the Chulmun period. Archaeologists have been arguing that rice became the hallmark of the ancient farmers' subsistence by the Songguk-Ri stage (ca. 2500 BP), and acted as a trigger of emergence of a social hierarchy and the subsequent social complexity (B. C. Kim, 2006a, 2006b). However, the results from Songguk-Ri in this study did not fully support the idea of rice as dominant subsistence strategy (cf. Kwak et al., 2017). At this point, I have to admit that the quantitative proportion of different food classes, which is derived from the analyses, cannot be assumed to be the direct representation of the diet taken by the ancient farmers. For example, even if the ratio of ruminants to porcine fat of a site is 1:3, this does not mean its ancient farmers relied exactly 3 times more on pork than ruminant animals. Therefore, it is reasonable that the results of the analyses in this thesis are viewed as a macroscopic explanation.

One surprising observation in this study was the absence of the presence of C_4 plants. Based on archaeobotanical analyses on both Chulmun and Mumun sites, G. Lee (Crawford & Lee, 2003; G. A. Lee, 2011) stressed the role of both broom corn and foxtail millet as major subsistence resources of the Mumun farmers. Since the study of G. Lee was based on well-organized systematic paleobotanical analyses, the presence of C_4 plant oil was highly anticipated in my analyses. Surprisingly, among the 113 samples collected from three Mumun sites, only one sample showed $\delta^{13}C$ values close to the C_4 range (KIM049: palmitic acid: -16.8 ‰, stearic acid: -17.2 ‰). However, it is yet to be confirmed whether this one sample indicates the presence of millet. Recent studies in China revealed that the $\delta^{13}C$ range of modern millet is from -10.48 to -10.05 ‰, higher than the average range of C_4 plants (-17 to -12.5 ‰) (cf. Malainey, 2011; Pechenkina et al., 2005).

There could be several explanations about the absence of C_4 plant oil. Firstly, it is possible that millet was processed/cooked without water; for example, popped or roasted (cf. Reber & Evershed, 2004a). Especially, considering the size of the millet grain which is smaller than that of the rice grain, popping might have been the major cooking method for millet, for it enlarges the size of the grain. Another possible explanation is that millet was often cooked with other food stuffs. One may assume that millet grains may have been cooked together with pork or other animals as an additive ingredient because of their small size.

If millet was cooked in a pot along with other resources, C16:0 and C18:0 fatty acids from this pot may indicate relatively low $\delta^{13}C$ values in comparison with those of millet. The result of the bulk isotope analysis on human remains from the Mumun period ($\delta^{13}C$: –12.2 ‰, $\delta^{15}N$: 10.1 ‰) might indicate this type of cooking method (D. I. Ahn, 2006, Figure 6.2). Relatively low $\delta^{13}C$ value in comparison with those of modern millet (from -10.48 to -10.05 ‰) and somewhat high $\delta^{15}N$ value indicate combination of millet, terrestrial mammal and marine resources.

Traditionally, the Chulmun – Mumun transition was explained as rapid, mainly due to climate-driven human migrations from the northeastern region of China (cf. J. S. Kim, 2003, 2006). Focusing on the overall differences in archaeological assemblages of the two periods, Korean archaeologists emphasized the discrepancy between the Chulmun and Mumun traditions. The Mumun migrants seem to be portrayed as a highly able group who could have eradicated the Chulmun indigenous foragers. Since the Mumun migrants were armed with new technology and an innovative subsistence strategy — intensive (rice) farming — they were able to spread suddenly and swiftly into the foraging contexts to leave little evidence of a transitional period (B. C. Kim, 2006a); and they were subsequently able to constrain the mobility of the indigenous hunter-gatherers by blocking their ways to resource patches to enhance the transition to farming (J. S. Kim, 2003, 2006). Rice eventually became the hallmark of the ancient farmers' subsistence by the middle Mumun period, and acted as a trigger of the emergence of a social hierarchy and the subsequent social complexity (B. C. Kim, 2006a, 2006b). After this somewhat expedient explanation was established, archaeologists tend to focused on finding farming tools and carbonized grains (especially, rice grains).

I have to admit that the results of organic geochemical analyses in this study do not mean harvested corps were never processed or cooked in a pot throughout the Mumun period. Animal (and fish) fats are much more concentrated than plant oil in general and the former often mask the signal of the latter, if they are cooked in a same pot. Therefore, I am not removing the possibility that some (or most) of the pots were used for cooking both animals and grains. Nevertheless, an overly simplified stress on rice farming blurs the real complexity of the Mumun subsistence (G. A. Lee, 2011). In many studies, the role of the other subsistence strategies in the Mumun farmers' diet, especially hunting terrestrial animals, have largely been neglected. However, the results of the organic geochemical analyses in this study showed that hunting and fishing persisted still well after rice farming was introduced.

6.2. The subsistence of Iron Age

In the Korean archaeology, the Iron Age is an area somewhat less studied using typical methods of archaeological science. This is partially because for the times since this period, vast documentary records from the Han Dynasty of China have been extensively employed for the interpretation of archaeological phenomena. Though these documents are valuable in terms of treating the contemporary past, they are neither chronicles nor meticulous ethnographies. They offer tantalizing snippets of information, allowing variable interpretations, as can be seen in the varying discussions about this period by Chinese and Korean historians (S. M. Nelson, 1993). Under these circumstances, not much information related to the subsistence of the Iron Age was released until the systematic paleobotanical investigation and isotope analysis regarding this period began to be effected (Choy & Richards, 2010; Y. J. Jeong, 2010; H. K. Lee, 2010).

According to the paleobotanical evidence given by various places in the central part of the Korean Peninsula (Y. J. Jeong, 2010; H. K. Lee, 2010), the Iron Age people had a diet focused on C_4 plants. They mainly consumed different kinds of millets (foxtail, broomcorn and Japanese millet). On the other hand, the isotopic evidence from the human bone collagen showed somewhat different possibilities. In 2009, Choy and Richards (2009) conducted the bulk carbon and nitrogen analysis on 48 human bones and 45 animal bones which were excavated from the Iron Age (ca. 200 BC – 100 AD) shell midden of Nuk-Do island, Sacheon city (cf. Y. N. Seo, 2004, Figure 2.7). The $\delta^{13}C$ and $\delta^{15}N$ values indicated that the people consumed C3 plants, terrestrial animals and possibly marine resources. Indeed, a direct comparison between the results of the investigations conducted in the inland and an island is not recommended, for the overall subsistence strategy might be quite different between the peoples of the two regions. Anyhow, we must remark that the paleobotanical evidence cannot provide any information related to animal consumption: a further investigation has to be conducted to see if terrestrial animals or aquatic resources were also regularly consumed at inland villages of the Iron Age.

According to the results of the organic geochemical analysis of the potsherds from the Eupha-Ri site (Figure 5.30; 5.31; 5.32; 5.35), most of the pots were used for processing animal resources. It strongly suggests the possibility that the Iron Age people regularly consumed animals such as ruminants and wild boars. Interestingly, no sample showed the presence of C_4 plant. Taking into account the study of Choy and Richards (2009), it is possible that C_4 plants were not the main part of the Iron Age diet. However, it is also possible that C_4 plants might have been cooked together with animals or potsherds from the pots used to cook C_4 plants such as millet were simply not sampled, for the samples were collected only from eight of the total 36 excavated house pits under the limited condition described in chapter five.

6.3. Luminescence dating

As I mentioned in chapter three, the luminescence dating is more effective than the other dating methods, especially in terms of pottery chronology. One of the two main goals of this thesis is to establish a long term chronology of subsistence from the Incipient/Early Mumun period to the Iron Age (3,400 - 2,000 BP). The radiocarbon dating does not date potteries themselves but the nearby organic remains (e.g. charcoal). This means the dating event inevitably has a variable relation to the target event of pottery manufacture. The luminescence dating reveals when the pot in question was made. In order to grasp the chronology of subsistence, archaeologists need to know the details of the cooking events. Since the cooking event is more likely to be associated with the manufacturing event than the depositional event, the luminescence dating is probably the most suitable method for establishing a subsistence chronology. I have to admit that the number of samples for the luminescence dating in this thesis is quite small to build a general chronology of subsistence from 3,400 to 2,000 BP. At the same time, I also have to admit that the sites included in this study are relatively well dated with AMS radiocarbon dating method. Since the charcoal samples for the radiocarbon dating were collected from the hearths inside of the house pits (cf. Figure 2.5), one may argue that the dates are relatively well associated with the cooking episodes. I do not disagree with this assumption and recognize the credibility of the published radiocarbon dates. However, on the other hand, I think we can still be benefited by the luminescence dating, due to its inherent nature of dating the manufacturing event.

	Luminescence dates (accumulative); calendar year	Radiocarbon dates (accumulative); calendar year
Sosa-Dong	1122 - 673 BC (2 dates)	1300 - 850 BC (16 dates)
Kimpo-Yangchon	808 - 310 BC (2 dates)	1420 - 415 BC (43 dates)
Songguk-Ri	N/A	835 - 200 BC (18 dates)
Eupha-Ri	373 BC - 594 AD (3 dates)	168 - 356 AD (4 dates)

Table 6.1: The comparison between the luminescence dates and AMS radiocarbon dates of the four sites

Table 6.1 shows the comparison between the radiocarbon dates and the luminescence dates of the four sites studied in this study. Note that all the dates were accumulated, including the error terms.

The biggest challenge in here is the relatively wide error terms in the luminescence dates (Table 5.4; 5.8; 5.14), in comparison with those in the radiocarbon dates (Table 5.1; 5.5; 5.9; 5.11). The average error term of the luminescence dates was about 120 years; and it created a range of error of about 240 years in each of the dates. Despite the relatively wider error terms of the luminescence dating compared to the Radiocarbon method, all the luminescence dates from the Mumun period were within the range of the Radiocarbon dates. The results from the Iron Age Eupha-Ri site was a bit different, indicating both the upper and lower limits of the luminescence dates exceeded those of the radiocarbon ones regardless of the error terms. This might be simply because the number of radiocarbon dates from the Eupha-Ri site is much smaller than those from the Mumun period sites, creating a narrower range of age. Despite the small sample size and the issue mentioned above, the luminescence and radiocarbon dates from the four sites are somewhat correlated with each other overall (Figure 6.4). Through the luminescence dating, at least in the macroscopic view, I was able to build a general chronology of subsistence between the sites.

6.4. Implications and future directions

Since the methods employed in this thesis are based on several assumptions, there are a number of potential sources of error in this study. Future directions will focus on overcoming those potential limitations.

First, boiling food inside of a pot is not the only option for cooking. For example, as I mentioned above, it is possible that millet was processed/cooked without water; for example, popped or roasted. Though ethnographic studies showed that boiling at a high temperature is regarded as a particularly effective cooking method in the preparation of faunal and floral resources in pots (Crown & Wills, 1995; Stahl, 1989; Wandsnider, 1997), strictly speaking, my study cannot reflect the entire subsistence change during the period in question. The most critical issue in here is whether there were more effective methods of cooking rice beside boiling in a pot. Rice is the center of existence in Asia, where more than 90 percent of the world's rice is grown. Traditionally, in East Asia (Korea, China, and Japan), the most familiar and well-known cooking method for rice is boiling with water (Luh, 1980). For special occasions, rice was used for making cakes, noodles or drinks (R. Barker, Herdt, & Rose, 1985). However, in terms of efficiency for the day to day consumption, boiling was the easiest way of cooking. Though we do not have solid evidence of how rice was cooked during the Mumun period, considering the known cooking methods, I argue that Mumun farmers probably preferred boiling. In addition, though it is reasonable to think that the pottery was mainly used as cooking vessels, rice might also have been cooked in bamboo tubes or other containers.

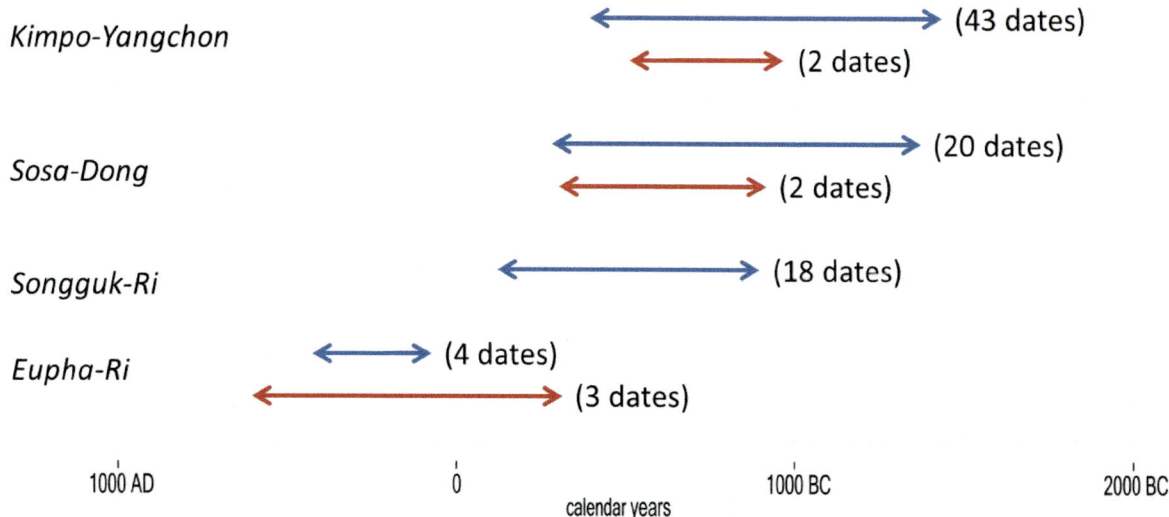

Figure 6.4: The comparison between the AMS radiocarbon dates and luminescence dates of the four sites

Second, the reference data used for CISA might not be suitable to the Korean Peninsula. Most of the reference data used in this thesis were generated by the modern wild fauna existing in Northern Europe (Copley et al., 2003; Craig et al., 2011; Dudd et al., 1999, 1998; Steele et al., 2010). In order to avoid the effects of commercial farming and selective breeding, the modern reference samples were collected from authentic wild animals. The question is whether the $\delta^{13}C$ values of the available modern samples from Northern Europe are comparable with those of the archaeological ones from the Korean Peninsula. Generally, in Europe, there have been only rare examples of wild C_4 plants (Tafuri et al., 2009). Therefore, wild herbivores mainly consumed C_3 plants in general. In case of the Korean Peninsula, studies showed that most of indigenous wild plants are C_3 plants; and the isotopic analysis revealed that the main food stuffs of wild animals are C_3 plants (D. I. Ahn, 2006; Choy & Richards, 2010; H. S. Kim, 2010; cf. J. J. Lee, 2011b). In these circumstances, one could argue that the basic environmental conditions which may affect the $\delta^{13}C$ values of the living organism in both regions are quite similar. However, I do admit that reference data used in this thesis might not perfectly reflect the environment in the Korean Peninsula. To avoid this potential problem, I have also employed Salque et al. (2013)'s approach which remove the exogenous factors related to the local environment (cf. Figure 5.33; 5.34; 5.35). Perhaps the best way to overcome this problem is to create CSIA reference data based on the indigenous fauna and flora from the Korean Peninsula. Future direction will focus on generating this reference.

Third, a pot is reused over time, and may be used to cook different kinds of food from one cooking episode to others. Since fatty acids and other compounds tend to accumulate in the fabric of the pot wall, the result of the organic geochemical analysis is more likely to reflect the entire usage of the pot. Also, because of the inherent nature of the isotope analysis, the result of CSIA is assumed to represent the type of food groups which were most frequently processed in it. This means that, even though the $\delta^{13}C$ values of a certain sample indicated the presence of porcine fat, and though I argue that the pot was mainly used for cooking pork, it is still possible that it was also used for cooking other food stuffs such as rice. In this regard, I also have to admit that the quantitative proportion of different food classes, which is derived from the analyses, cannot be assumed to be the direct representation of the diet taken by the ancient farmers of the site in question. Therefore, it is reasonable that the results of this thesis are viewed as a macroscopic explanation. Nevertheless, I think they have enough quality to give an insight into the human subsistence of the ancient Korea and the role of the intensive rice agriculture in the prehistoric Korean diet.

Fourth, it is possible that a certain type of pottery (or certain pot) was used for a certain type of food. As I mentioned in Chapter two, there are some variations in patterns on the Mumun potteries (Figure 2.4). In this setting, a pottery with a certain pattern might have been used for cooking rice. It is also possible that the frequency of usage might be different from one pot to another. For example, what if people ate rice for twenty meals a week, and other food for one meal a week, cooking the one and the other respectively in the same pot and in a number of different pots? In this case, linking the numbers of sampled sherds directly to the overall diet might not be proper for understanding the true nature of the subsistence. One way to overcome this limitation is to analyze sherds with different patterns and see if we can observe any trend. We can also investigate whether the frequency of usage can be distinguished by the organic geochemical analysis through the experiment with laboratory cooking episodes (e.g. one time vs. 10 times or more). Future researches will include these approaches.

Fifth, the study does not include the subsistence pattern of the Late Mumun period. Though not many studies have been conducted in relation to the subsistence of the Late Mumun period, it is assumed that rice was the main food stuff (cf. J. J. Lee, 2011b). To make a more convincing argument about the overall subsistence of the Mumun period, for future researches, I need to conduct the organic geochemical analysis on potsherds from the Late Mumun period village sites to see if I can observe a dramatic increase in rice consumption. Also, more Iron Age sites have to be included in the future researches. As I mentioned above, according to the paleobotanical evidence given by various places in the central part of the Korean Peninsula (Y. J. Jeong, 2010; H. K. Lee, 2010), the Iron Age people had a diet focused on C_4 plants. They mainly consumed different kinds of millets (foxtail, broomcorn and Japanese millet). On the other hand, the isotopic evidence from shell middens indicates that they consumed C_3 plants, terrestrial animals and possibly marine resources (Choy & Richards, 2009). Though one Iron Age site, Eupha-Ri, was included in this thesis, the result was not enough to properly address the subsistence pattern during the period. Therefore, for better understanding, it is critical to analyze more potsherds from Iron Age villages.

Sixth, the study might have overlooked the role of domesticated animals in the ancient Korean farmers' diet. In many cases, the places where we can observe the evidence of domesticated plants also tend to show that of domesticated animals. This is because the harvested crops from agriculture can be easily used for provisioning the livestock. Since domesticated plants may show a carbon isotope signal different from that of available indigenous wild plants, there is a strong possibility that domesticated animals show different $\delta^{13}C$ values, compared to the wild ones. The reference $\delta^{13}C$ value ranges I employed in this study were based on the data from wild animals, assuming the role of domesticated animals in the ancient Korean farmer's diet is minimal at least until the Mumun period. Unfortunately, the domestication of animals is one of the least-studied areas in the Korean archaeology, mainly due to the high acidity of the sediments which does not allow long-term preservation of organisms. Among the available terrestrial mammals considered as major subsistence resources in the prehistoric Korean Peninsula, the strongest candidate for domestication is the pig. Though cattle, horse and dog were also considered, the main purpose of their domestication was not consumption. According to recent carbon and nitrogen isotope analyses on pig bones excavated from shell middens of the Korean Peninsula (J. J. Lee, 2011a), the isotopic signal of the pig shows it become omnivorous from the historic proto Kingdom period (ca. AD 0 - 250). Wild boars are herbivores in general (J. J. Lee, 2011a), and both the $\delta^{13}C$ and $\delta^{15}N$ signal of pig bones from the Chulmun and Mumun periods were quite low (Figure 6.1; 6.2), indicating they mainly consumed wild C_3 plants. These results support my assumption about the domesticated animals in the ancient Korean farmers' diet. Overall, I think domesticated animals may have played a little role in subsistence throughout the prehistoric periods in Korea.

Lastly, the sampling of potsherds might have distorted the true nature of the subsistence pattern. Ideally, the entire pottery from each site need to be analyzed. However, due to the restriction in funding, sampling was inevitable. On one hand, since I wanted to maximize the representation of the entire site, I had to include the entire house pits that yield potteries. On the other hand, I had to compromise with myself in fixing the number of samples per one house pit, for the budget for the analyses was limited. In this regard, I admit that my sampling strategy might have distorted the real picture (for example, it is possible that we did not observe strong evidence of rice simply because the pots used to cook rice were not sampled). However, at the same time, I also think that this possibility is quite low, for I was able to observe a similar subsistence pattern at all the four sites included in this study.

7. Conclusion

7.1. Reprising the work so far

In chapter one, I briefly reviewed recent approaches to understanding the subsistence change from foragers to farmers. Several underlying characteristics that the places showing the evidence of farming have in common were also mentioned. These characteristics include (1) opportunistic migrations of small groups of people, (2) ambiguity in the results of the genetic studies, and (3) selective adaptation of new subsistence strategy. Then, I elucidated the main purpose of this thesis by narrowing down the region investigated to the central part of the Korean Peninsula and addressing the current ideas that Korean archaeologists have on the role of the intensive rice agriculture among them.

In chapter two, firstly, the general history and social context of Korean archaeology from its beginning stage to the present time was stated, focusing especially on the political upheavals such as the Japanese annexation of the country and the Korean War. Then, I showed the cultural historical background of the two main prehistoric periods in question: the Chulmun and Mumun periods. After that, I discussed current views on the transition from foragers to farmers in the Korean Peninsula and its main problematic assumption: strict dichotomy between Chulmun hunter-gatherers and Mumun full-dress rice farmers. Most importantly, the central hypothesis of this thesis—that there was utilization of a wide range of animal and plant resources along with rice among the ancient farmers in the central part of the Korean Peninsula—was proposed based on the recent scientific evidence from Korea and Europe. Lastly, I re-evaluated the current rice-base model and its problematic assumption of the strict dichotomy between Chulmun hunter-gatherers and Mumun full-dress rice farmers by presenting the possibility of continuity between the Chulmun and the Mumun periods. I also tried to interpret the dichotomy/continuity between the Chulmun and the Mumun periods with the concept of essentialism versus materialism.

Chapters three and four were dedicated respectively to the two main analytical methods of this study: luminescence dating and organic geochemical analyses. In these chapters, I discussed the methods, research design and analytical procedure of the luminescence dating and organic geochemical analysis. I briefly outlined the history of the two methods employed in the discipline of archaeology, elucidated some of their main principles and emphasized why these two methods are essential to achieve my goal. Also, some of the important implications related to the methods were listed. Lastly, the details of the laboratory experimental procedures were elucidated.

In the fifth chapter, I presented the results of the luminescence dating the organic geochemical analyses on the four inland habitation sites (Sosa-Dong, Kimpo-Yangchon, Songguk-Ri, Eupha-Ri). Firstly, I did an in-depth review of the overall archaeological records of the four sites. Then, the sampling strategy, methods and the results of the organic geochemical analyses and luminescence dating for each of the sites were elucidated one by one. For the interpretation of the data that were produced by the two methods, available archaeological records and radiocarbon dates were incorporated, because this type of scientific research can be strengthen by the proper archaeological contextual information.

In chapter six, I further examined the initial interpretations that I had made in the former chapter with the available bulk isotope and paleobotanical data. First, by correlating the results of the organic geochemical analyses and luminescence dating with the available bulk isotope and paleobotanical data, I argued hunting and fishing continued after the introduction of rice farming and the role of intensive farming in the prehistoric Korean peninsula was over estimated. Lastly, I listed some of the important implications of the results in this study.

7.2. Transition from foraging to farming: Theoretical model vs. Empirical world

Now I have the results in hand, I will return to the issue raised in chapter one of what the most important factors are in explaining the emergence of agriculture. As I mentioned in chapter one, recent approaches to understanding the subsistence change from foragers to farmers could fall into four categories: (1) population pressure, (2) climatic fluctuation, (3) cultural or social processes, and (4) evolutionary processes. The population pressure model, climatic fluctuation model, and evolutionary model usually assume external stresses and emphasize the capacity of farming as a stress reliever. In these models, people use agriculture as a risk-reduction strategy against resource stress driven by environmental changes. Recently, in Korean archaeology, there is a heated debate over the evidence of external stresses (e.g. population increase or sea level change) around 4,000–3,000 BP and the introduction of rice farming as a stress reliever (cf. K. D. Bae et al., 2013; J. C. Kim & C. J. Bae, 2010; J. S. Kim, 2003, 2006; G. A. Lee, 2011; J. J. Lee, 2001). However, the suggested evidence of resource stress driven by environmental change is still limited and requires further investigation.

Above all, in the Korean Peninsula, if rice agriculture was used as a risk-reduction strategy, rice should be the mainstay of the Korean diet from the Mumun period. However, according to the results of the organic geochemical analyses on the potsherds from major habitation sites such as Sosa-Dong, Kimpo-Yangchon and Songguk-Ri, hunted wild animals were a significant part of the Mumun farmers' day to day foodstuffs. This means the Mumun subsistence pattern does not fully support those models based on the external stresses. The cultural or social model does not convincingly explain the Mumun subsistence either, for the evidence of conceptual ideas such as a new cosmology, religious practices, symbolic behaviors or a wide range of information in relation to rice farming is not clear.

The subsistence pattern we can reveal from the results of the organic geochemical analyses in this study and the limited (or controversial) evidence related to the resource stress suggest that the Korean Peninsula might have been a relatively stable/rich resource zone by the time of the transition from foraging to farming. Interestingly, in East Asia, we do observe similar patterns. For example, the Yangtze River Valley corridor in China, where we can find the earliest evidence of rice agriculture in the world, was a very rich resource zone (Silva et al., 2015; B. D. Smith, 1998, 2007). Also, Jomon Japan, the period that is traditionally considered as giving an affluent hunter-gathering context, showed solid evidence of plant domestication (Obata et al., 2007). In none of these areas did domestication of plants and agriculture appear to have developed within a necessity is the mother of invention context (B. D. Smith, 2007: p. 197). Then, why can we observe the initiation of agriculture in these rich resource zones?

This resource rich context for the initial domestication of plants and development of agriculture fits with the expectations of the niche-construction theory (Laland & Brown, 2006; Laland et al., 2001; Odling-Smee et al., 2003). The niche construction is defined as organism-driven environmental modification and activities of organisms that bring about changes in environment. From this point of view, not only does an environment cause changes in species through selection, but species also cause changes in their environment through niche construction. This means, within this theoretical framework, it is possible that an organism can actively change its environment for their own purposes without experiencing other causal factors. Humans are acknowledged to be the ultimate niche constructors, both in terms of the diversity of different ways in which we manipulate the environment around us for our own benefits and the magnitude of our resultant impacts (B. D. Smith, 2007: p. 195). In this perspective, agriculture can be one of the acmes of human niche construction. The evidence from Japan and China showed that domestication of plants and agriculture developed within rich resource conditions that enable the continuous human experimental intervention in the environment. According to niche-construction theory, the rich resource zones, those that exhibited a greater capacity for supporting more people in more permanent settlements, could be expected to have witnessed stronger sustained niche-construction efforts (cf. Odling-Smee et al., 2003). The wider the range of species included in human efforts of intervention became, and the more different potential forms of intervention human could attempt, the greater the likelihood of domestication and agriculture would have been successful (B. D. Smith, 2007).

Then, what was the motivation for the ancient Koreans to create their own niche through practicing rice agriculture? As I mentioned in Chapter two, it was the increased sedentism (Price & Gebauer, 1995: p. 8). The evidence from Japan and China also showed that before the initial domestication of plants and development of agriculture, hunter-gatherers started to take the sedentary life style (Crawford, 2011; B. D. Smith, 1998, 2007). According to Bar-Yosef (Rocek & Bar-Yosef, 1998) the non-agricultural sedentism requires storage technologies and containers such as pottery or special pits in which to store food securely. Also, it requires sufficient year-round, easily accessible local natural resources.

In Korean archaeology, there is a heated debate over the evidence of sedentism around the final stage of the Neolithic period (Ahn et al, 2015; Hah, 2016). However, at least in some part of the peninsula, we do observe solid evidence of a long-term, permanent occupation of the peninsula by complex hunter-gatherers at various places since around 6,000 BP. At the Amsa-Dong Site in south-east Seoul (Figure 2.7), at least 12 houses, a significant amount of pottery and different types of ground stone tools such as arrow points, spear points and sickles, were excavated (Im, 1985). The house structures and seasonality of the faunal assemblages at the Tongsam-Dong site (Figure 2.7) in the southern part of the Korean Peninsula indicate that people lived there year-round on a permanent basis (G. A. Lee, 2011; J. J. Lee, 2001). In this sedentary life style, along with hunting and gathering, prehistoric Koreans already had specific subsistence solutions which included distinctive combinations of wild (e.g. acorn, Manchurian walnut), possibly managed (e.g. chenopod, panicoid grass), and domesticated (e.g. foxtail and broomcorn millet, possibly soybean, azuki and beefsteak plant) plants from 5,500 BP (G. A. Lee, 2011: p. S326). The prehistoric Koreans created their niche long before the initiation of rice agriculture. Rice agriculture was just an another addition of environment engineering (niche construction) to get a more reliable resource.

The transitions from foragers to farmers that occurred around the world had various and diverse pathways and probably cannot be fully explained with a few generalized models. This diversity motivates us to investigate the specific manifestations of this transition in different parts of the world and better understand the different ways that people made this profound transformation. In the central part of the prehistoric Korean Peninsula, from the beginning of the Mumun period (c.a. 3,400 BP) we observe the solid evidence of the intensive rice farming. However, even after rice farming was introduced, people still relied on hunting and gathering of both terrestrial animals and marine resources. In the central part of the Korean Peninsula, the indigenous foragers adopted new subsistence strategies little by little for their own purposes (cf. Crawford, 2011; G. A. Lee, 2011; Robb, 2013; B. D. Smith, 1995, 2007, 2011).

7.3. Concluding remark: The role of intensive agriculture as a subsistence strategy in the prehistoric Korean peninsula

In this study I focused on the four inland habitation sites (Sosa-Dong, Kimpo-Yangchon, Songguk-Ri, Eupha-Ri) in the central part of the Korean Peninsula, a region that contains a vast amount of archaeological materials related to the subsistence change in the deep past. The aim of this research was re-evaluating the conventional rice-centered models to better understand the overall pattern of subsistence strategies and assess the weight of rice in it. To achieve this goal the study tested the central hypothesis that a wide range of resources were utilized along with rice between 3,400 and 2,000 BP. The results of the organic geochemical analyses on the potsherds from the four sites supported the suggested hypothesis, indicating that most of the pots were used for processing terrestrial animals and marine resources.

In the central Korean Peninsula, past efforts to reconstruct the ancient dietary patterns have been challenged by the high acidity of the sediments (RDA 1988). Because of these acidic sediments, the direct examination of the remains of subsistence resources in the Korean Peninsula is limited to relatively special locations that provide better preservation of bone or plant remains such as caves, rock-shelters, or shell middens (cf. Choy & Richards, 2009, 2010; Choy et al., 2012).

In terms of archaeological records, it is clear that the intensive agriculture was practiced in the central part of the Korean Peninsula as early as around 3,400 BP (G. A. Lee, 2003, 2011). Solid evidence of dry fields, irrigated rice paddies and harvesting tools have been found (Yoon & Bae, 2010). During this period, many large scale inland villages started to appear. However, most of these sites did not yield paleobotanical evidence and faunal remains due to post-depositional processes and the high acidity of the archaeological sediments. Due to these conditions, Korean archaeologists are not able to recover detailed information about the diet of the ancient Korean farmers, and the main focus has been put on harvested crops such as rice (B. C. Kim, 2006b; cf. G. A. Lee, 2011).

The subsistence change related to the emergence of agriculture always has been the critical part of anthropological debates. This subsistence change has been often described as a transition from hunter gathering to intensive agriculture. However, in many areas, scholars have only focused on how quickly or completely people abandoned wild terrestrial and marine resources after the introduction of domesticated plants (cf. Craig et al., 2011). Once the domesticated plants are introduced, the role of other food resources in the ancient farmers' diet is neglected. In the Korean archaeology, rice has been often considered as a dominant subsistence resource since 3,400 BP. The possibility of other subsistence strategies in those farmers' diet, for example, hunting terrestrial animals and procuring aquatic resources, were largely undermined. However, the results of the organic geochemical analyses and luminescence dating suggest that both terrestrial and aquatic animals were a considerable part of the ancient farmers' diet, well after farming was introduced. Though It is unquestionable that the intensive rice agriculture was practiced in the Korean Peninsula as early as 3,400 BP, the results of this study indicated that there was a wider range of resource utilization along with (rice) farming.

Bibliography

Adams, W. Y. (2001). Classification and typology. In Smelser, N. J. & Baltes, P. B. (eds.) *International encyclopedia of the social and behavioral sciences*, 1962–1966. Elsevier.

Adams, W. Y., & Adams, E. W. (2007). *Archaeological typology and practical reality: A dialectical approach to artifact classification and sorting*. Cambridge University Press.

Ahn, D. I. (2006). Dietary reconstruction by stable isotopic analysis: The Konam-ri shell midden in Korea. *Journal of the Korean Ancient Historical Society* 54, 5–20. In Korea.

Ahn, J. H. (1991). *A study of Mumun pottery from southern Korea*. PhD dissertation, Kyungpook National University. In Korean.

Ahn, J. H. (2000). Organization of Korean agricultural society. *Journal of Korean Archaeological Society* 43, 41–66. In Korean.

Ahn, J. H. (2001). Characterizing Middle Mumun pottery period settlements. *Yongnam Archaeological Review* 29, 1–42. In Korean.

Ahn, S. M. (2004). *The beginning of agriculture and sedentary life and their relation to social changes in Korea*. Cultural diversity and the archaeology of the 21st century - The society of archaeological studies 50th anniversary symposium at Okayama, Japan.

Ahn, S. M., Kim, J. S. & Hwang, J. H. (2015). Sedentism, settlements and radiocarbon dates of Neolithic Korea. *Asian Perspective* 54(1), 111–143.

Aikens, C. M., Ames, K. M. & Sanger, D. (1986). Evolutionary ecology and the social sciences. In Akazawa, C and Aikens, C. M. (eds.) *Prehistoric hunter-gatherers in Japan: New research methods*, 3–26. University of Tokyo Press.

Aitken, M. J. (1985). *Thermoluminescence dating*. London: Academic Press.

Aitken, M. J. (1998). *An introduction to optical dating: The dating of Quaternary sediments by the use of photon-stimulated luminescence*. Oxford university Press.

Bae, J. S. (2007). *The establishment of Mumun pottery period and complex society*. Seoul: Seokyeong Press. In Korean.

Bae, K. D., Bae, C. J. & Kim, J. C. (2013). Reconstructing human subsistence strategies during the Korean Neolithic: Contributions from zooarchaeology, geosciences, and radiocarbon dating. *Radiocarbon* 55(23), 1350–1357.

Bale, M. T. (2012). *Storage practices, intensive agriculture, and social change in Mumun pottery period Korea, 2903 - 2450 calibrated years BP*. PhD dissertation, University of Toronto.

Banerjee, D., Murray, A. S., Bøtter-Jensen, L. & Lang, A. (2001). Equivalent dose estimation using a single aliquot of polymineral fine grains. *Radiation Measurements* 33(1), 73–94.

Bang, J. H., Kim, K. D. & Eum, C. H. (2009). Age comparisons of coastal sand dune stratum in Chollipo, Korea by altering preheat and cut-heat, and grain size distributions by OSL dating. *Analytical Science and Technology* 22(1), 51–56.

Bar-Yosef, O. (2011). Climatic fluctuations and early farming in west and East Asia. *Current Anthropology* 52(S4), S175–S193.

Barker, A., Venables, B., Stevens Jr, S. M., Seeley, K. W., Wang, P. & Wolverton, S. (2012). An optimized approach for protein residue extraction and identification from ceramics after cooking. *Journal of Archaeological Method and Theory* 19(3), 407–439.

Barker, R., Herdt, R. W. & Rose, B. (1985). *The rice economy of Asia*. Washington, D. C.: Resource for the Future.

Barton, L., Newsome, S. D., Chen, F. H., Wang, H., Guilderson, T. P. & Bettinger, R. L. (2009). Agricultural origins and the isotopic identity of domestication in northern China. *Proceedings of the National Academy of Sciences* 106(14), 5523–5528.

Belfer-Cohen, A. & Goring-Morris, A. N. (2011). Becoming farmers. *Current Anthropology* 52(S4), S209–S220.

Bentley, R. A., Tayles, N., Higham, C., Macpherson, C. & Atkinson, T. C. (2007). Shifting gender relations at Khok Phanom Di, Thailand. *Current Anthropology* 48(2), 301–314.

Berstan, R., Dudd, S. N., Copley, M. S., Morgan, E. D., Quye, A. & Evershed, R. P. (2004). Characterisation of 'bog butter' using a combination of molecular and isotopic techniques. *Analyst* 129(3), 270–275.

Bethell, P. H., Goad, L. J., Evershed, R. P., & Ottaway, J. (1994). The study of molecular markers of human activity: The use of coprostanol in the soil as an indicator of human faecal material. *Journal of Archaeological Science* 21(5), 619–632.

Bettinger, R. L., Barton, L., Richerson, P. J., Boyd, R., Wang, H. & Choi, W. (2007). The transition to agriculture in northwestern China. *Developments in Quaternary Sciences* 9, 83–101.

Binford, L. R. (1968). Post-Pleistocene adaptations. In Binford, S R. & Binford, L. R. (eds.), *New perspectives in archaeology*. Chicago: Aldine Publishing Company.

Bleed, P. & Matsui, A. (2010). Why didn't agriculture develop in Japan? A consideration of Jomon ecological style, niche construction, and the origins of domestication. *Journal of Archaeological Method and Theory* 17(4), 356–370.

Boland, J. M. (1990). Leapfrog migration in North American shorebirds: Intra-and interspecific examples. *The Condor* 92(2), 284–290.

Bonde, A., Murray, A. & Friedrich, W. L. (2001). Santorini: Luminescence dating of a volcanic province using quartz? *Quaternary Science Reviews* 20(5), 789–793.

Boone, J. L. & Smith, E. A. (1998). Is it evolution yet? A critique of evolutionary archaeology 1. *Current Anthropology* 39(S1), S141–S174.

Borić, D. (2002). The Lepenski Vir conundrum: Reinterpretation of the Mesolithic and Neolithic sequences in the Danube Gorges. *Antiquity* 76(294), 1026–1039.

Boutton, T. W. (1991). Stable carbon isotope ratios of natural materials: II. atmospheric, terrestrial, marine, and freshwater environments. In Colman, D. C. & Fry, B. (eds.) *Carbon Isotope Techniques*, 173–185. San Diego: Academic Press.

Bramanti, B., Thomas, M. G., Haak, W., Unterlaender, M., Jores, P., Tambets, K., ... others. (2009). Genetic discontinuity between local hunter-gatherers and central Europe's first farmers. *Science* 326(5949), 137–140.

Bull, I. D., van Bergen, P. F., Bol, R., Brown, S., Gledhill, A. R., Gray, A. J., ... Evershed, R. P. (1999). Estimating the contribution of Spartina anglica biomass to salt-marsh sediments using compound specific stable carbon isotope measurements. *Organic Geochemistry* 30(7), 477–483.

Buonasera, T. Y., Tremayne, A. H., Darwent, C. M., Eerkens, J. W. & Mason, O. K. (2015). Lipid biomarkers and compound specific δ 13 c analysis indicate early development of a dual-economic system for the arctic small tool tradition in Northern Alaska. *Journal of Archaeological Science* 61, 129–138.

Buyeo National Museum. (2000). *Songguk-ri VI*. Buyeo: Buyeo National Museum.

Calvin, M., & Benson, A. A. (1948). The path of carbon in photosynthesis. *Science* 107(2784), 476–480.

Cannon, M. D. & Broughton, J. M. (2010). Evolutionary ecology and archaeology: An introduction. In Broughton, J. M. & Cannon, M. D. (eds.), *Evolutionary ecology and archaeology: Applications to problems in human evolution and prehistory*, 1–12. University of Utah Press.

Cauvin, J. (1994). *Naissance des divinités, naissance de l'agriculture: La révolution des symboles au Néolithique*. Paris: CNRS.

Cavalli-Sforza, L. L., Menozzi, P., & Piazza, A. (1994). *The history and geography of human genes*. Princeton university Press.

Charters, S., Evershed, R. P., Goad, L. J., Leyden, A., Blinkhorn, P. W., & Denham, V. (1993). Quantification and distribution of lipid in archaeological ceramics: Implications for sampling potsherds for organic residue analysis and the classification of vessel use. *Archaeometry* 35(2), 211–223.

Cheon, S. H. (2005). Formation and development of Doldae Mun pottery in Korea. *Journal of the Korean Archaeological Society* 57, 61–97. In Korean.

Childe, V. G. (1951). *Man makes himself*. New York: New American Library.

Chilliard, Y., Ferlay, A. & Doreau, M. (2001). Effect of different types of forages, animal fat or marine oils in cow's diet on milk fat secretion and composition, especially conjugated linoleic acid (CLA) and polyunsaturated fatty acids. *Livestock Production Science* 70(1), 31–48.

Choi, J. H., Murray, A. S., Cheong, C. S., Hong, D. G., & Chang, H. W. (2006). Estimation of equivalent dose using quartz isothermal TL and the SAR procedure. *Quaternary Geochronology* 1(2), 101–108.

Choy, K. C. & Richards, M. P. (2009). Stable isotope evidence of human diet at the Nukdo shell midden site, South Korea. *Journal of Archaeological Science* 36(7), 1312–1318.

Choy, K. C. & Richards, M. P. (2010). Isotopic evidence for diet in the Middle Chulmun period: A case study from the Tongsamdong shell midden, Korea. *Archaeological and Anthropological Sciences* 2(1), 1–10.

Choy, K. C., Ahn, D. M. & Richards, M. P. (2012). Stable isotopic analysis of human and faunal remains from the Incipient Chulmun (Neolithic) shell midden site of Ando Island, Korea. *Journal of Archaeological Science* 39(7), 2091–2097.

Cohen, D. J. (2003). Microblades, pottery, and the nature and chronology of the Palaeolithic-Neolithic transition in China. *The Review of Archaeology* 24(2), 21–36.

Cohen, D. J. (2011). The beginnings of agriculture in China. *Current Anthropology* 52(S4), S273–S293.

Cohen, M. N. (1977). *The food crisis in prehistory: Overpopulation and the origins of agriculture*. New Haven: Yale University Press.

Cohen, M. N. (2009). Introduction: Rethinking the origins of agriculture. *Current Anthropology* 50(5), 591–595.

Condamin, J., Formenti, F., Metais, M. O., Michel, M. & Blond, P. (1976). The application of gas chromatography to the tracing of oil in ancient amphorae. *Archaeometry* 18(2), 195–201.

Copley, M. S., Berstan, R., Dudd, S. N., Docherty, G., Mukherjee, A. J., Straker, V., ... Evershed, R. P. (2003). Direct chemical evidence for widespread dairying in prehistoric Britain. *Proceedings of the National Academy of Sciences* 100(4), 1524–1529.

Copley, M. S., Berstan, R., Dudd, S. N., Straker, V., Payne, S. & Evershed, R. P. (2005a). Dairying in antiquity. I. evidence from absorbed lipid residues dating to the British Iron Age. *Journal of Archaeological Science* 32(4), 485–503.

Copley, M. S., Berstan, R., Dudd, S., Aillaud, S., Mukherjee, A. J., Straker, V., ... Evershed, R. P. (2005b). Processing of milk products in pottery vessels through British prehistory. *Antiquity* 79(306), 895–908.

Copley, M. S., Bland, H. A., Rose, P., Horton, M., & Evershed, R.P. (2005c). Gas chromatographic, mass spectrometric and stable carbon isotopic investigations of organic residues of plant oils and animal fats employed as illuminants in archaeological lamps from Egypt. *Analyst* 130(6), 860–871.

Copley, M. S., Rose, P. J., Clapham, A., Edwards, D. N., Horton, M. C. & Evershed, R. P. (2001). Detection of palm fruit lipids in archaeological pottery from Qasr Ibrim, Egyptian Nubia. Proceedings of the Royal Society of London. *Series B: Biological Sciences* 268(1467), 593–597.

Corr, L. T., Richards, M. P., Jim, S., Ambrose, S. H., Mackie, A., Beattie, O. & Evershed, R. P. (2008). Probing dietary change of the Kwäday Dän Ts' ìnchi individual, an ancient glacier body from British Columbia: I. complementary use of marine lipid biomarker and carbon isotope signatures as novel indicators of a marine diet. *Journal of Archaeological Science* 35(8), 2102–2110.

Correa-Ascencio, M. & Evershed, R. P. (2014). High throughput screening of organic residues in archaeological potsherds using direct acidified methanol extraction. *Analytical Methods* 6(5), 1330–1340.

Craig, O. E., Love, G. D., Isaksson, S., Taylor, G. & Snape, C. E. (2004). Stable carbon isotopic characterization of free and bound lipid constituents of archaeological ceramic vessels released by solvent extraction, alkaline hydrolysis and catalytic hydropyrolysis. *Journal of Analytical and Applied Pyrolysis* 71(2), 613–634.

Craig, O. E., Saul, H., Lucquin, A., Nishida, Y., Taché, K., Clarke, L., ... others. (2013). Earliest evidence for the use of pottery. *Nature* 496(7445), 351–354.

Craig, O. E., Steele, V. J., Fischer, A., Hartz, S., Andersen, S. H., Donohoe, P., ... others. (2011). Ancient lipids reveal continuity in culinary practices across the transition to agriculture in Northern Europe. *Proceedings of the National Academy of Sciences* 108(44), 17910–17915.

Cramp, L. J. E., Evershed, R. P. & Eckardt, H. (2011). What was a mortarium used for? Organic residues and cultural change in Iron Age and Roman Britain. *Antiquity* 85(330), 1339–1352.

Crawford, G. W. (2008). The Jomon in early agriculture discourse: Issues arising from Matsui, Kanehara and Pearson. *World Archaeology* 40(4), 445–465.

Crawford, G. W. (2009). Agricultural origins in North China pushed back to the Pleistocene – Holocene boundary. *Proceedings of the National Academy of Sciences* 106(18), 7271–7272

Crawford, G. W. (2011). Advances in understanding early agriculture in Japan. *Current Anthropology* 52(S4), S331–S345.

Crawford, G. W. & Lee, G. A. (2003). Agricultural origins in the Korean Peninsula. *Antiquity* 77(295), 87–95.

Crown, P. L. & Wills, W. H. (1995). The origins of southwestern ceramic containers: Women's time allocation and economic intensification. *Journal of Anthropological Research* 51(2), 173–186.

Dudd, S. N. & Evershed, R. P. (1998). Direct demonstration of milk as an element of archaeological economies. *Science* 282(5393), 1478–1481.

Dudd, S. N., Evershed, R. P. & Gibson, A. M. (1999). Evidence for varying patterns of exploitation of animal products in different prehistoric pottery traditions based on lipids preserved in surface and absorbed residues. *Journal of Archaeological Science* 26(12), 1473–1482.

Dudd, S. N., Regert, M. & Evershed, R. P. (1998). Assessing microbial lipid contributions during laboratory degradations of fats and oils and pure triacylglycerols absorbed in ceramic potsherds. *Organic Geochemistry* 29(5), 1345–1354.

Duller, G. A. T. (2008). *Luminescence dating: Guidelines on using luminescence dating in archaeology.* Swindon: English Heritage.

Dunne, J., Evershed, R.P., Salque, M., Cramp, L., Bruni, S., Ryan, K., Biagetti, S., & di Lernia, S. (2012) First dairying in green Saharan Africa in the fifth millennium BC. *Nature* 486, 390-394.

Dunnell, R. C. (1970). Seriation method and its evaluation. *American Antiquity* 35(3), 305–319.

Dunnell, R. C. (1971). *Systematics in prehistory.* New York: Free Press.

Earle, T. K. (2002). *Bronze Age economics: The beginnings of political economies.* Westview Press.

Edwards, C. J., MacHugh, D. E., Dobney, K. M., Martin, L., Russell, N., Horwitz, L. K., ... others. (2004). Ancient DNA analysis of 101 cattle remains: Limits and prospects. *Journal of Archaeological Science* 31(6), 695–710.

Eerkens, J. W. (2001). *The origins of pottery among late prehistoric hunter-gatherers in California and the Western Great Basin*, PhD dissertation. University of California Santa Barbara.

Eerkens, J. W. (2002). The preservation and identification of Piñon resins by GC-MS in pottery from the Western Great Basin. *Archaeometry* 44(1), 95–105.

Eerkens, J. W. (2005). GC - MS analysis and fatty acid ratios of archaeological potsherds from the Western Great Basin of North America. *Archaeometry* 47(1), 83–102.

Eerkens, J. W. (2007). Organic residue analysis and the decomposition of fatty acids in ancient potsherds. *BAR International Series* 1650, 90.

Ehleringer, J. R., Lin, Z. F., Field, C. B., Sun, G. C. & Kuo, C. Y. (1987). Leaf carbon isotope ratios of plants from a subtropical monsoon forest. *Oecologia* 72(1), 109–114.

Enser, M. (1991). Animal carcass fats and fish oils. In Pritchard, J. L. & Rossell, J. B. (eds.) *Analysis of oilseeds, fats and fatty foods*, 329–394) Elsevier Applied Science.

Evershed, R. P. (1993). Biomolecular archaeology and lipids. *World Archaeology* 25(1), 74–93.

Evershed, R. P. (2007). Exploiting molecular and isotopic signals at the Mesolithic-Neolithic transition. In Whittle, A. & Cummings, V. (eds.) *Going over: The Mesolithic-Neolithic transition in north-west Europe.* Proceedings of the British Academy, London: British Academy.

Evershed, R. P. (2008a). Experimental approaches to the interpretation of absorbed organic residues in archaeological ceramics. *World Archaeology* 40(1), 26–47.

Evershed, R. P. (2008b). Organic residue analysis in archaeology: The archaeological biomarker revolution. *Archaeometry* 50(6), 895–924.

Evershed, R. P., Arnot, K. I., Collister, J., Eglinton, G., & Charters, S. (1994). Application of isotope ratio monitoring gas chromatography - mass spectrometry to the analysis of organic residues of archaeological origin. *Analyst* 119(5), 909–914.

Evershed, R. P., Bethell, P. H., Reynolds, P. J. & Walsh, N. J. (1997). 5β-stigmastanol and related 5β-stanols as biomarkers of manuring: Analysis of modern experimental material and assessment of the archaeological potential. *Journal of Archaeological Science* 24(6), 485–495.

Evershed, R. P., Copley, M. S., Dickson, L. & Hansel, F. A. (2008). Experimental evidence for the processing of marine animal products and other commodities containing polyunsaturated fatty acids in pottery vessels. *Archaeometry* 50(1), 101–113.

Evershed, R. P., Dudd, S. N., Anderson-Stojanovic, V. R., & Gebhard, E. R. (2003). New chemical evidence for the use of combed ware pottery vessels as beehives in ancient greece. *Journal of Archaeological Science* 30(1), 1–12.

Evershed, R. P., Dudd, S. N., Charters, S., Mottram, H. R., Stott, A. W., Raven, A., ... Bland, H. A. (1999). Lipids as carriers of anthropogenic signals from prehistory. *Philosophical Transactions of the Royal Society B: Biological Sciences* 354(1379), 19–31.

Evershed, R. P., Dudd, S. N., Copley, M. S., Berstan, R., Stott, A. W., Mottram, H. R., ... Crossman, Z. (2002). Chemistry of archaeological animal fats. *Accounts of Chemical Research 35*(8), 660–668.

Evershed, R. P., Dudd, S. N., Lockheart, M. J. & Jim, S. (2001). Lipids in archaeology. In Brothwell, D. R. & Pollard, A. M. (eds.) *Handbook of archaeological sciences*, 331–349. Willey.

Evershed, R. P., Heron, C. & John Goad, L. (1990). Analysis of organic residues of archaeological origin by high-temperature gas chromatography and gas chromatography-mass spectrometry. *Analyst* 115(10), 1339–1342.

Fankhauser, B. (1997). Amino acid analysis of food residues in pottery: A field and laboratory study. *Archaeology in Oceania* 32(1), 131–140.

Feathers, J. K. (2003). Use of luminescence dating in archaeology. *Measurement Science and Technology* 14(9), 1493–1509.

Feathers, J. K. (2009). Problems of ceramic chronology in the Southeast: Does shell-tempered pottery appear earlier than we think? *American Antiquity* 74(1), 113–142.

Flannery, K. V. (1972). The cultural evolution of civilizations. *Annual Review of Ecology and Systematics* 3, 399–426.

Flannery, K. V. (1976). *The early Mesoamerican village*. Orlando: Academic Press.

Fraser, S. E., Insoll, T., Thompson, A. & van Dongen, B. E. (2012). Organic geochemical analysis of archaeological medicine pots from Northern Ghana. the multi-functionality of pottery. *Journal of Archaeological Science* 39(7), 2506–2514.

Friedli, H., Lötscher, H., Oeschger, H., Siegenthaler, U. & Stauffer, B. (1986). Ice core record of the $^{13}C/^{12}C$ ratio of atmospheric CO_2 in the past two centuries. *Nature* 324(6094), 237–238.

Fuller, D. Q., Qin, L., Zheng, Y., Zhao, Z., Chen, X., Hosoya, L. A. & Sun, G. P. (2009). The domestication process and domestication rate in rice: Spikelet bases from the Lower Yangtze. *Science* 323(5921), 1607–1610.

Fuller, D. Q., Sato, Y., Castillo, C., Qin, L., Weisskopf, A. R., Kingwell-Banham, E. J., … others. (2010). Consilience of genetics and archaeobotany in the entangled history of rice. *Archaeological and Anthropological Sciences* 2(2), 115–131.

Galili, E., Rosen, B., Gopher, A. & Kolska-Horwitz, L. (2003). The emergence and dispersion of the eastern Mediterranean fishing village: Evidence from submerged Neolithic settlements off the Carmel coast, Israel. *Journal of Mediterranean Archaeology* 15(2), 167–198.

Gremillion, K. J. & Piperno, D. R. (2009). Human behavioral ecology, phenotypic (developmental) plasticity, and agricultural origins. *Current Anthropology* 50(5), 615–619.

Hah, I. S. (2016). Consideration on the Late Neolithic – Early Bronze Age transition. *Proceedings of the 2016 joint conference of the Society for Korean Bronze Culture and Korean Neolithic Research Society*. Daejeon: Chungnam National University..

Han, C. G., & Son, G. E. (2000). The layers and artifacts of the Sorori Paleolithic site (Area B) in Cheongwon. *Practical Sciences* 14, 635–653. In Korean.

Hansel, F. A., Copley, M. S., Madureira, L. A. & Evershed, R. P. (2004). Thermally produced ω-(oalkylphenyl) alkanoic acids provide evidence for the processing of marine products in archaeological pottery vessels. *Tetrahedron Letters* 45(14), 2999–3002.

Hastorf, C. A., & DeNiro, M. J. (1985). Reconstruction of prehistoric plant production and cooking practices by a new isotopic method. *Nature* 315, 489–491.

Heron, C. and Craig, O. E. (2015). Aquatic resources in foodcrusts: identification and implication. *Radiocarbon* 57(4), 707–719.

Heron, C., & Evershed, R. P. (1993). The analysis of organic residues and the study of pottery use. *Archaeological Method and Theory* 5, 247–284.

Heron, C., Evershed, R. P. & Goad, L. J. (1991). Effects of migration of soil lipids on organic residues associated with buried potsherds. *Journal of Archaeological Science* 18(6), 641–659.

Ho, S. Y. W., Larson, G., Edwards, C. J., Heupink, T. H., Lakin, K. E., Holland, P. W. H. & Shapiro, B. (2008). Correlating bayesian date estimates with climatic events and domestication using a bovine case study. *Biology Letters* 4(4), 370–374.

Hong, D. G., Yi, S. B., Galloway, R. B. & Tsuboi, T. (2001). Optical dating of archaeological samples using a single aliquot of quartz stimulated by blue light. *Journal of Radioanalytical and Nuclear Chemistry* 247(1), 179–184.

Hong, D. S., Galloway, R., Kim, M. & Park, S. (2003). Optical dating of the hydroponic farm at Korea. *Journal of Radioanalytical and Nuclear Chemistry* 256(2), 365–368.

Huntley, D. J., Godfrey-Smith, D. I. & Thewalt, M. L. W. (1985). Optical dating of sediments. *Nature* 313, 105–107.

Im, H. J. (1985). *Amsadong -a Neolithic village site on the Han river-*. Seoul: Seoul National University Museum.

Jansen, T., Forster, P., Levine, M. A., Oelke, H., Hurles, M., Renfrew, C., ... Olek, K. (2002). Mitochondrial DNA and the origins of the domestic horse. *Proceedings of the National Academy of Sciences* 99(16), 10905–10910.

Jeong, Y. J. (2010). Rice cultivation during the Proto Three Kingdoms period in Korea. *Journal of the Korean Ancient Historical Society* 69, 19–38. In Korean.

Kim, B. C. (2006a). Political economy of wet rice cultivation in the Bronze period in central western Korea. *Journal of the Korean Archaeological Society* 58, 40–65. In Korean.

Kim, B. C. (2006b). Political versus subsistence economy of Songgukri culture in Chungnam province. *Journal of Honam Archaeological Society* 24, 65–96. In Korean.

Kim, B. M., Kim, A. G., Kang, B. H., Chae, A. R., Cho, U. J., Oh, K. T. & Park, S. N. (2008). *Sosa-dong*. Seoul: The Korea Institute of Heritage.

Kim, B. M., Soh, S. Y., Lee, J. A., Kang, T. H., Koo, J. M., Mah, W. Y. & Kim, M. J. (2013). *Kimpo-Yangchon*. Hanam: The Korea Institute of Heritage.

Kim, G. T., Seo, H. J., Jeong, C. Y. & Joo, H. M. (2011). *Songguk-ri XII, XIII*. Buyeo: The Korean National University of Cultural Heritage.

Kim, G. T., Seo, H. J., Jeong, C. Y. & Joo, H. M. (2013). *Songguk-ri XIV*. Buyeo: The Korean National University of Cultural Heritage.

Kim, H. S. (2010). The age and dietary concerned about human remain of Daepo shell midden. *Journal of Korean Neolithic Society* 20, 89–111. In Korean.

Kim, J. C. & Bae, C. J. (2010). Radiocarbon dates documenting the Neolithic-Bronze Age transition in Korea. *Radiocarbon* 52(3), 483–492.

Kim, J. C., Duller, G. A. T., Roberts, H. M., Wintle, A. G., Lee, Y. I. & Yi, S. B. (2010). Reevaluation of the chronology of the palaeolithic site at Jeongokri, Korea, using OSL and TT-OSL signals from quartz. *Quaternary Geochronology* 5(2), 365–370.

Kim, J. C., Duller, G. A. T., Roberts, H. M., Wintle, A. G., Lee, Y. I. & Yi, S. B. (2009). Dose dependence of thermally transferred optically stimulated luminescence signals in quartz. *Radiation Measurements* 44(2), 132–143.

Kim, J. S. (2001). Reconsidering the Heunamri assemblages: Origins and date. *Yongnam Archaeological Review* 28, 35–64. In Korean.

Kim, J. S. (2003). Land-use conflict and the rate of the transition to agricultural economy: A comparative study of southern Scandinavia and central-western Korea. *Journal of Archaeological Method and Theory* 10(3), 277–321.

Kim, J. S. (2006). Resource patch sharing among foragers: Lack of territoriality or strategic choice. In Grier, C., Kim. J. S. & Uchiyama, J. (eds.) *Beyond affluent foragers: Rethinking hunter-gatherer complexity*, 168–191. Oxford: Oxbow press.

Kim, M. J., Park, M. S., Lee, S. J., Nah, H. R. & Hong, H. W. (2012). *Optical dating of potteries excavated from Pungnabtoseong earthen wall, Seoul*. Transactions of the Korean Nuclear Society Spring Meeting. In Korean.

Kim, M. K. (2008). Multivocality, multifaceted voices, and Korean archaeology. In Habu, J. & Matsunaga, J. M. (eds.) *Evaluating multiple narratives*, 118–137. Springer.

Kim, M. K. (2015). Rice in ancient Korea: Status symbol or community food? *Antiquity* 89(346), 838–853.

Kim, S. H. (2004). Rice is ethnic spirit. *Segye Ilbo* June 7, P.27. In Korean.

Kitcher, P. (1981). Explanatory unification. *Philosophy of Science* 48(4), 507–531.

Kitcher, P. (1989). Explanatory unification and the causal structure of the world. *Scientific Explanation* 13, 410–505.

Kobayashi, T., Kaner, S., & Nakamura, O. (2004). *Jomon reflections: Forager life and culture in the prehistoric Japanese archipelago.* Oxford: Oxbow Books.

Kolattukudy, P. E. (1976). *Chemistry and biochemistry of natural waxes.* Amsterdam: Elsevier.

Koyama, S., Thomas, D. H. & Hakubutsukan, K. M. (1981). *Affluent foragers: Pacific coast East and West.* Osaka: National Museum of Ethnology.

Kwak, B. C. (2005). Soro-Ri rice. *The Hankyoreh* December 9, p.26. In Korean.

Kwak, S. K., Kim, G. T. & Lee, G. A. (2017). Beyond rice farming: Evidence from central Korea reveals wide resource utilization in the Songgukri culture during the late-Holocene. *Holocene* (published online).

Laland, K. N. & Brown, G. R. (2006). Niche construction, human behavior, and the adaptive-lag hypothesis. *Evolutionary Anthropology: Issues, News, and Reviews* 15(3), 95–104.

Laland, K. N., Odling-Smee, F. J. & Feldman, M. W. (2001). Cultural niche construction and human evolution. *Journal of Evolutionary Biology* 14(1), 22–33.

Lee, B. G. (1974). Mumun pottery and polished stone tools of Gyeonggi-do. Kogohak 3, 53–129. In Korean.

Lee, G. A. (2003). *Changes in subsistence patterns from the Chulmun to Mumun periods: Archaeobotanical investigation.* PhD dissertation, University of Toronto.

Lee, G. A. (2011). The transition from foraging to farming in prehistoric Korea. *Current Anthropology* 52(S4), S307–S329.

Lee, H. K. (2010). A study on the establishment of the crop assemblage of the Proto-Three Kingdoms period in central Korea. *Journal of Korean Archaeological Society* 75, 98–125. In Korean.

Lee, H. W. (2008). The upper and lower limits of Early Bronze Age in South Korea. *Journal of Korean Bronze Culture* 1, 28–63.

Lee, J. J. (2001). *From shellfish gathering to agriculture in prehistoric Korea: The Chulmun to Mumun transition.* PhD dissertation, University of Wisconsin-Madison.

Lee, J. J. (2006). Analysis of the shellfish from the Youngyudo and Ulwangdong Islands: Review of the site function. In Yi, S. B., Lim, S., Yang, S. & Hong, E. (eds.) *Excavation report at nambuk-dong and eulwang-dong i site*, 269–280. Seoul National University Museum. In Korean.

Lee, J. J. (2009). Archaeological approaches to documenting animals in the Korean Peninsula. In Ahn, S. M. & Lee, J. J. (eds.) *New approaches to prehistoric agriculture*, 252–269. Seoul: Sahoi Pyoungnon. In Korean.

Lee, J. J. (2011a). Domesticated pig in Korea: Its socioeconomic and symbolic context. *Journal of Korean Archaeological Society* 79, 131–174. In Korean.

Lee, J. J. (2011b). Intensification of millet and rice agriculture in Korea - evidence from stable isotopes -. *Journal of the Korean Ancient Historical Society* 73, 31–66. In Korean.

Lee, S. H. (2005). *A study of Geomdan-ri type pottery*. PhD dissertation, Busan National University. In Korean.

Lee, Y. J. & Woo, J. Y. (2002). *The excavation of the Paleolithic Sorori rice and its important problems*. First international conference of Cheongwon county - prehistoric agriculture in Asia and Sorori rice. Cheongju: Chungbuk National University Museum.

Lim, H. S., Chung, C. H., Kim, C. B., Lee, Y. I., Lee, H. J. & Lee, Y. C. (2007). Late-Holocene palaeoclimatic change at the Dongnimdong archaeological site, Gwangju, SW Korea. *Holocene* 17(5), 665–672.

Liu, L., Lee, G. A., Jiang, L. & Zhang, J. (2007). The earliest rice domestication in China. *Antiquity* 81(313).

Lowe, T. E., Peachey, B. M. & Devine, C. E. (2002). The effect of nutritional supplements on growth rate, stress responsiveness, muscle glycogen and meat tenderness in pastoral lambs. *Meat Science* 62(4), 391–397.

Lu, T. L. (2006). The occurrence of cereal cultivation in China. *Asian Perspectives* 45(2), 129–158.

Luh, B. S. (1980). *Rice: Production and utilization*. AVI Publishing Co., Inc.

Malainey, M. E. (2011). *A consumer's guide to archaeological science: Analytical techniques*. New York: Springer Science & Business Media.

Malhi, R. S., Kemp, B. M., Eshleman, J. A., Cybulski, J., Smith, D. G., Cousins, S. & Harry, H. (2007). Mitochondrial haplogroup M discovered in prehistoric North Americans. *Journal of Archaeological Science* 34(4), 642–648.

Malmström, H., Gilbert, M. T. P., Thomas, M. G., Brandström, M., Stortextbackslasha a, J., Molnar, P., ... others. (2009). Ancient DNA reveals lack of continuity between Neolithic hunter-gatherers and contemporary Scandinavians. *Current Biology* 19(20), 1758–1762.

Marwick, B. (2008). Beyond typologies: The reduction thesis and its implications for lithic assemblages in Southeast Asia. *Bulletin of the Indo-Pacific Prehistory Association* 28, 108–116.

Mayr, E. (1959). Darwin and the evolutionary theory in biology. In Meggers, B. J. (ed.) *Evolution and anthropology: A centennial appraisal*, 1–10. Washington, D. C.: Anthropological Society of Washington.

McCormac, F. G., Baillie, M. G. L., Pilcher, J. R., Brown, D. M. & Hoper, S. T. (1994). $\delta^{13}C$ measurements from the Irish oak chronology. *Radiocarbon* 36(1), 27–35.

McKeever, S. W. S. & Chen, R. (1997). Luminescence models. *Radiation Measurements* 27(5), 625–661.

Medina, E. & Minchin, P. (1980). Stratification of δ13C values of leaves in Amazonian rain forests. *Oecologia* 45(3), 377–378.

Medina, E., Klinge, H., Jordan, C. & Herrera, R. (1980). Soil respiration in amazonian rain forests in the Rio-Negro-Basin. *Flora* 170(3), 240–250.

Milner, N., Craig, O. E., Bailey, G. N., Pedersen, K. & Andersen, S. H. (2004). Something fishy in the Neolithic? A re-evaluation of stable isotope analysis of Mesolithic and Neolithic coastal populations. *Antiquity* 78(299), 9–22.

Mirabaud, S., Rolando, C. & Regert, M. (2007). Molecular criteria for discriminating adipose fat and milk from different species by nanoESI MS and MS/MS of their triacylglycerols: Application to archaeological remains. *Analytical Chemistry* 79(16), 6182–6192.

Morton, J. D. & Schwarcz, H. P. (1988). Stable isotope analysis of food residue from Ontario ceramics. In *Proceedings of the 26th international archaeometry symposium: Held at university of Toronto, Canada,* 89–93. University of Toronto.

Mottram, H. R., Dudd, S. N., Lawrence, G. J., Stott, A. W. & Evershed, R. P. (1999). New chromatographic, mass spectrometric and stable isotope approaches to the classification of degraded animal fats preserved in archaeological pottery. *Journal of Chromatography A* 833(2), 209–221.

Mukherjee, A. J. (2004). *The importance of pigs in the later British Neolithic: Integrating stable isotope evidence from lipid residues in archaeological potsherds, animal bone, and modern animal tissues,* PhD dissertation. University of Bristol.

Murray, A. S. & Wintle, A. G. (2000). Luminescence dating of quartz using an improved single-aliquot regenerative-dose protocol. *Radiation Measurements* 32(1), 57–73.

National Museum of Korea. (1979). *Songguk-ri I.* Seoul: National Museum of Korea.

National Museum of Korea. (1986). *Songguk-ri II.* Seoul: National Museum of Korea.

National Museum of Korea. (1987). *Songguk-ri III.* Seoul: National Museum of Korea.

National Research Institute of Cultural Heritage. (2002). *Dictionary of Korean archaeology.* Daejeon: National Research Institute of Cultural Heritage.

Nelson, S. M. (1993). The archaeology of Korea. Cambridge: Cambridge University Press.

Nishida, M. (1983). The emergence of food production in Neolithic Japan. *Journal of Anthropological Archaeology* 2(4), 305–322.

Nishitani, T. (2014). *The history of exchages between the ancient Japan and Korean Peninsula.* Tokyo: Dohsei Press. In Japanese.

Noh, T. D. (2003). The new insight on the history of exchange between ancient China and Korea. *Journal of Korean Ancient History* 32, 5–11. In Korean.

Norton, C. J. (2000). Subsistence change at Konam-ri: Implications for the advent of rice agriculture in Korea. *Journal of Anthropological Research* 56(3), 325–348.

Norton, C. J. (2007). Sedentism, territorial circumscription, and the increased use of plant domesticates across Neolithic-Bronze Age Korea. *Asian Perspectives* 46(1), 133–165.

Obata, H., Sasaki, Y. & Senba, Y. (2007). Impressions on pottery revealed cultivation of Glycine max subsp. max (soybean) in the late to latest jomon preriods in Kyushu Island. *Japanese Journal of Historical Botany* 15, 97–114.

O'Brien, M. J., & Lyman, R. L. (1998). *James a. Ford and the growth of Americanist archaeology.* University of Missouri Press.

Odling-Smee, F. J., Laland, K. N. & Feldman, M. W. (2003). *Niche construction: The neglected process in evolution.* Princeton University Press.

Patrick, M., Koning, A. J., & Smith, A. B. (1985). Gas liquid chromatographic analysis of fatty acids in food residues from ceramics found in the Southwestern Cape, South Africa. *Archaeometry* 27(2), 231–236.

Pechenkina, E. A., Ambrose, S. H., Xiaolin, M. & Benfer, R. A. (2005). Reconstructing northern Chinese Neolithic subsistence practices by isotopic analysis. *Journal of Archaeological Science* 32(8), 1176–1189.

Prescott, J. R., Huntley, D. J., & Hutton, J. T. (1993). Estimation of equivalent dose in thermoluminescence dating - the Australian slide method. *Ancient TL* 11(1), 1–5.

Prescott, J. R. & Hutton, J. T. (1994). Cosmic ray contributions to dose rates for luminescence and ESR dating: Large depths and long-term time variations. *Radiation Measurements* 23(2-3), 497-500.

Preusser, F., Degering, D., Fuchs, M., Hilgers, A., Kadereit, A., Klasen, N., ... Spencer, J. Q. G. (2008). Luminescence dating: Basics, methods and applications. *Quaternary Science Journal* 57(1-2), 95–149.

Price, T. D. (1995). Social inequality at the origins of agriculture. In Price, T. D. & Feinman, G. M. (eds.), *Foundations of social inequality*, 129–151. Springer.

Price, T. D. & Bar-Yosef, O. (2011). The origins of agriculture: New data, new ideas. *Current Anthropology* 52(S4), S163–S174.

Price, T. D. & Gebauer, A. B. (1995). *Last hunters, first farmers: New perspectives on the prehistoric transition to agriculture*. School of American Research Press.

Rafferty, J. (2008). Settlement patterns, occupations, and field methods. In Ratterty, J. & Peacock, E. (eds.) Time's river: Archaeological syntheses from the lower mississippi river valley, 99–124. Tuscaloosa: University of Alabama Press.

Reber, E. A. & Evershed, R. P. (2004a). How did Mississippians prepare maize? The application of compound-specific carbon isotope analysis to absorbed pottery residues from several Mississippi Valley sites. *Archaeometry* 46(1), 19–33.

Reber, E. A. & Evershed, R. P. (2004b). Identification of maize in absorbed organic residues: A cautionary tale. *Journal of Archaeological Science* 31(4), 399–410.

Redman, C. L. (1978). *The rise of civilization: From early farmers to urban society in the ancient Near East*. San Francisco: WH Freeman.

Regert, M., Vacher, S., Moulherat, C. & Decavallas, O. (2003). Adhesive production and pottery function during the Iron Age at the site of Grand Aunay (Sarthe, France). *Archaeometry* 45(1), 101–120.

Reimer, P. J., Bard, E., Bayliss, A., Beck, J. W., Blackwell, P. G., Bronk Ramsey, C., ... others. (2013). IntCal13 and marine13 radiocarbon age calibration curves 0-50,000 years cal BP. *Radiocarbon* 55(4), 1869–1887.

Reynard, L. M. & Hedges, R. E. M. (2008). Stable hydrogen isotopes of bone collagen in palaeodietary and palaeoenvironmental reconstruction. *Journal of Archaeological Science* 35(7), 1934–1942.

Richards, M. P., Pearson, J. A., Molleson, T. I., Russell, N. & Martin, L. (2003). Stable isotope evidence of diet at Neolithic Çatalhöyük, Turkey. *Journal of Archaeological Science* 30(1), 67–76.

Rindos, D. (1984). *The origins of agriculture: An evolutionary perspective*. Academic Press.

Robb, J. (2013). Material culture, landscapes of action, and emergent causation: A new model for the origins of the European Neolithic. *Current Anthropology* 54(6), 657–683.

Roberts, H. M. & Wintle, A. G. (2001). Equivalent dose determinations for polymineralic fine-grains using the SAR protocol: Application to a Holocene sequence of the Chinese loess plateau. *Quaternary Science Reviews* 20(5), 859–863.

Robertson, I., Switsur, V. R., Carter, A. H. C., Barker, A. C., Waterhouse, J. S., Briffa, K. R. & Jones, P. D. (1997). Signal strength and climate relationships in $^{13}C/^{12}C$ ratios of tree ring cellulose from oak in east England. *Journal of Geophysical Research: Atmospheres* 102(D16), 19507–19516.

Rocek, T. R. & Bar-Yosef, O. (1998). *Seasonality and sedentism: Archaeological perspectives from Old and New World sites* (Vol. 6). Cambridge: Peabody Museum Press.

Rowley-Conwy, P. (2009). Human prehistory: Hunting for the earliest farmers. *Current Biology* 19(20), R948–949.

Salque, M., Bogucki, P. I., Pyzel, J., Sobkowiak-Tabaka, I., Grygiel, R., Szmyt, M. & Evershed, R. P. (2013). Earliest evidence for cheese making in the sixth millennium BC in Northern Europe. *Nature* 493(7433), 522–525.

Seo, Y. N. (2004). *The shell mound and graves on the Neuk-do site.* Busan: Busan National University museum.

Seo, Y. S. (1981). A primary study on the history of Sino-Korean relations in the ancient period. *Academic Mongraphs* 5, 35–48. In Korean.

Shelach, G. (2000). The earliest Neolithic cultures of Northeast China: Recent discoveries and new perspectives on the beginning of agriculture. *Journal of World Prehistory* 14(4), 363–413.

Shin, G.Y., Kang, D.Y., Kim, S.H. & Jung, Y. D. (2013). Isotopic dietary history of Neolithic people from Janghang site at Gadeok Island, Busan. *Analytical science* 26(6), 387–394. In Korean.

Shin, S. J. (2007). Retrospect and prospect of the research on the Korean Neolithic culture. In *Issues in Neolithic culture in central-western Korea: Proceedings of the 2007 joint conference of the seoul-gyeonggi archaeological society and Korean Neolithic society*, 7–38. Seoul: National Museum of Korea.

Shoda, S. (2008). The traces of cooking on Bronze Age potteries. In Foodways Society (eds.) *The archaeology of cooking*, 37–45. Seoul: Seokyeong Press.

Silva, F., Stevens, C. J., Weisskopf, A., Castillo, C., Qin, L., Bevan, A. & Fuller, D. Q. (2015). Modelling the geographical origin of rice cultivation in Asia using the rice archaeological database. *PloS One* 10(9), e0137024.

Smith, B. D. (1989). Origins of agriculture in eastern North America. *Science* 246(4937), 1566–1571.

Smith, B. D. (1998). *The emergence of agriculture.* New York: W.H. Freeman.

Smith, B. D. (2007). Niche construction and the behavioral context of plant and animal domestication. *Evolutionary Anthropology: Issues, News, and Reviews* 16(5), 188–199.

Smith, B. D. (2011). The cultural context of plant domestication in eastern North America. *Current Anthropology* 52(S4), S471–S484.

Stahl, A. B. (1989). Plant-food processing: Implications for dietary quality. In Harris, D. R. & Jillman, G. C. (eds.), *Foraging and farming: The evolution of plant exploitation*, 171–194. London: Unwin Hyman.

Stear, N. A. (2008). *Changing patterns of animal exploitation in the Prehistoric Eurasian steppe: An integrated molecular, stable isotopic and archaeological approach.* PhD dissertation, University of Bristol.

Steele, V. J., Stern, B. & Stott, A. W. (2010). Olive oil or lard?: Distinguishing plant oils from animal fats in the archeological record of the eastern Mediterranean using gas chromatography/combustion/isotope ratio mass spectrometry. *Rapid Communications in Mass Spectrometry* 24(23), 3478–3484

Steward, J. H. (1972). *Theory of culture change: The methodology of multilinear evolution*. University of Illinois Press.

Strevens, M. (2004). The causal and unification approaches to explanation unified - causally. *Noûs* 38(1), 154–176.

Tafuri, M. A., Craig, O. E. & Canci, A. (2009). Stable isotope evidence for the consumption of millet and other plants in Bronze Age Italy. *American Journal of Physical Anthropology* 139(2), 146–153.

Thompson, A. H., Chaix, L. & Richards, M. P. (2008). Stable isotopes and diet at ancient Kerma, Upper Nubia (Sudan). *Journal of Archaeological Science* 35(2), 376–387.

Trigger, B. G. (1984). Alternative archaeologies: Nationalist, colonialist, imperialist. *Man* 19(3), 355–370.

Trigger, B. G. (2008). Alternative archaeologies in historical perspective. In Habu, J. & Matsunaga, J. M. (eds.) *Evaluating multiple narratives*, 187–195. Springer.

Tsukamoto, S., Murray, A. S., Huot, S., Watanuki, T., Denby, P. M. & Bøtter-Jensen, L. (2007). Luminescence property of volcanic quartz and the use of red isothermal TL for dating tephras. *Radiation Measurements* 42(2), 190–197.

Vilà, C., Leonard, J. A., Götherström, A., Marklund, S., Sandberg, K., Lidén, K., … Ellegren, H. (2001). Widespread origins of domestic horse lineages. *Science* 291(5503), 474–477.

Vogel, J. C. (1978). Recycling of CO_2 in a forest environment. *Oecologia Plantarum* 13, 89–94.

Wandsnider, L. (1997). The roasted and the boiled: Food composition and heat treatment with special emphasis on pit-hearth cooking. *Journal of Anthropological Archaeology* 16(1), 1–48.

Wang, H. J., Gong, S. J., Choi, T. S., Kim, I. G., Seo, S. D. & Kim, H. R. (2013). *Excavation report of the potential Hoengseong leisure park ground plot*. Wonju: Yonsei University Wonju museum.

Welch, P. D. & Scarry, C. M. (1995). Status-related variation in foodways in the Moundville chiefdom. *American Antiquity* 60(3), 397–419.

Whittle, A., and Cummings, V. (2007). *Going over: the Mesolithic-Neolithic transition in Western Europe*. Proceedings of the British Academy. London: British Academy.

Winterhalder, B. & Kennett, D. J. (2006). Behavioral ecology and the transition from hunting and gathering to agriculture. In Kennett, D. J. & Winterhalder, B. (eds.), *Behavioral ecology and the transition to agriculture*, 1–21. University of California press.

Winterhalder, B. & Kennett, D. J. (2009). Four neglected concepts with a role to play in explaining the origins of agriculture. *Current Anthropology* 50(5), 645–648.

Winterhalder, B., & Smith, E. A. (1992). Evolutionary ecology and the social sciences. In Smith, A. D. & Winterhalder, B. (eds.) *Evolutionary ecology and human behavior*, 3–23. Transaction Publishers.

Wintle, A. G. (1973). Anomalous fading of thermo-luminescence in mineral samples. *Nature* 245, 143–144.

Wintle, A. G., & Murray, A. S. (2006). A review of quartz optically stimulated luminescence characteristics and their relevance in single-aliquot regeneration dating protocols. *Radiation Measurements* 41(4), 369–391.

Wylie, A. (2002). *Thinking from things: Essays in the philosophy of archaeology*. Berkeley: University of California Press.

Woo, J. Y. (2012). Seoul food: Treating your idol to lunch is the true test of fandom. *Wall Street Journal: Life and Style*. February 23.

Yoon, T. Y., & Bae, J. S. (2010): *Agrarian society and the leader*. Seoul: National museum of Korea.

Zhao, Z. (2011). New archaeobotanic data for the study of the origins of agriculture in China. *Current Anthropology* 52(S4), S295–S306.